ZERO

ZERO

The Starting Point of Creating Reality

Chunsia

First Sentient Publications edition 2025

First published in Korean under the title: Zero 현실을 창조하는 마음상태
by 천시아 지음 © 2012 by Zenbook, Seoul, Korea

A paperback original

Korean cover design by KUSH | English cover adaptation by Laura Waltje
Korean book design by 김태민 | English book design by Laura Waltje

---

Library of Congress Control Number: 2024946353

Publisher's Cataloging-in-Publication Data

Names: Youn-kyung, Jun, 1985, author.
Title: Zero : the starting point of creating reality / Chunsia.
Description: Includes bibliographical references and index. | Boulder, CO: Sentient Publications, 2025.
Identifiers: LCCN: 2024946353 | ISBN: 978-1-59181-334-7 (paperback) | 978-1-59181-335-4 (ebook)
Subjects: LCSH Self actualization (Psychology) | Conduct of life. | Metaphysics. | Spirituality. | Conciousness. | Self help. | BISAC BODY, MIND & SPIRIT / New Thought | SELF-HELP / Personal Growth / Success | PHILOSOPHY / Metaphysics | SCIENCE / Physics / Quantum Theory | RELIGION / Spirituality
Classification: LCC BF161.C48 2025 | DDC 158.1--dc23

SENTIENT PUBLICATIONS
A Limited Liability Company
PO Box 1851
Boulder, CO 80306
www.sentientpublications.com

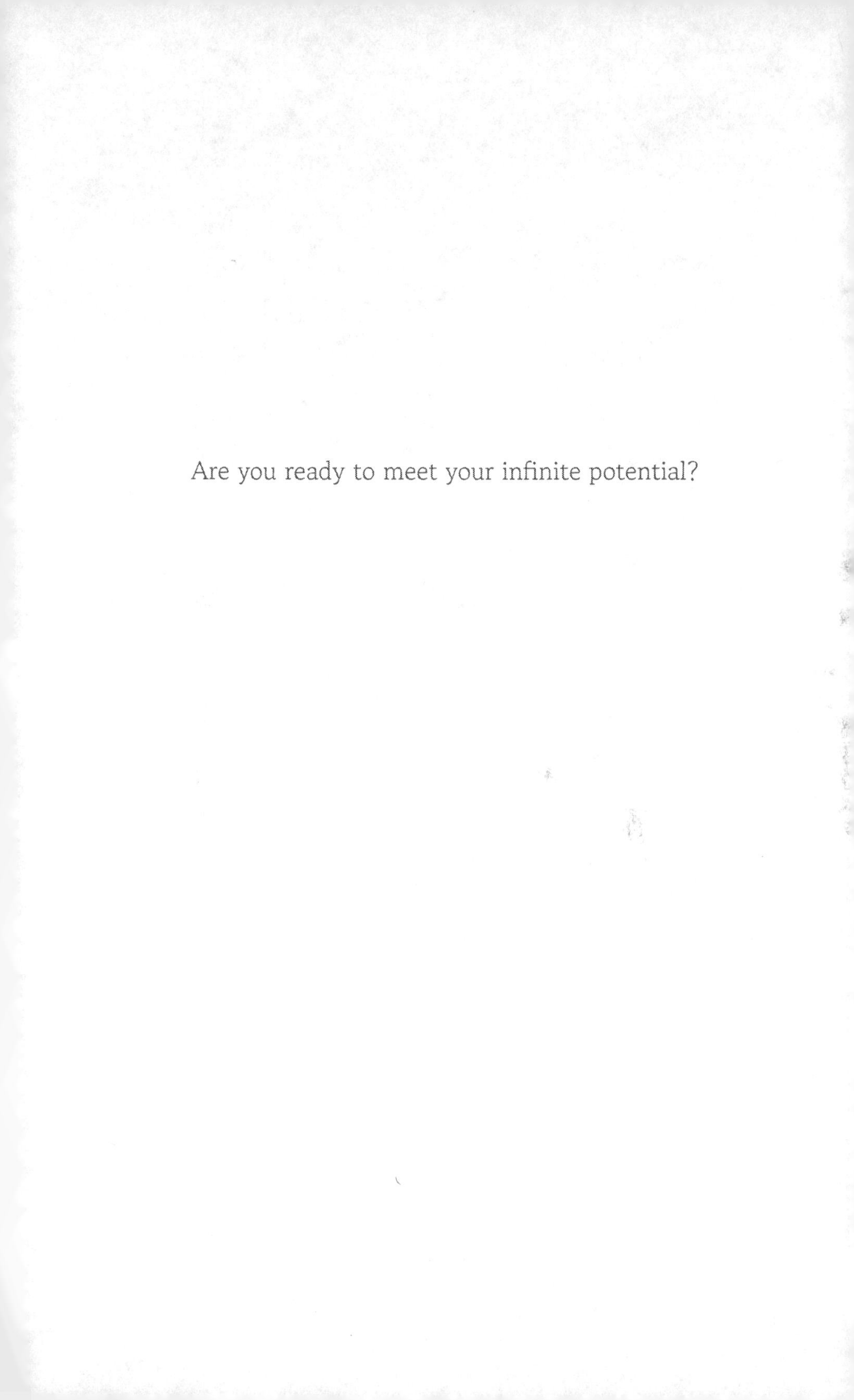

Are you ready to meet your infinite potential?

Contents

Foreword

Many years ago I interviewed the Fourteenth Dalai Lama for an article published in the Associated Press. His Holiness had recently publicly mentioned that he would probably be the last Dalai Lama. In that he is the fourteenth in the string of incarnations, each of which is believed to be the reincarnation of the former in a process dating back over six hundred years, I presented the question, "If you do not come back as the Dalai Lama, what will you do? You will be out of a job."

The Dalai Lama laughed and mischievously replied, "Perhaps I will come back as someone's pet dog, or even as a book. Or as an incognito North American like you."

Although said in humor, the Dalai Lama's words echoed the thoughts of a fourth century Indian Buddhist master by the name of Asanga. In one of his treatises, Asanga comments that, when a bodhisattva achieves enlightenment, his stream of being transforms into the *Trikāya* or "three bodies:" Mind into the *dharmakāya*, or formless gnosis dimension; Speech into the bliss body; and Body into physical manifestations. Asanga goes on to state that the physical manifestations can be anything useful in a direct way to specific beings. The pet dog, for example, would relieve the loneliness of its owner; and a book could provide life-changing inspiration for those lost and confused. In other words, a book can be a beacon of light to those lost in a dark world.

Chunsia's book *Zero* is a bit like that. It is part spiritual diary, and part "tips from lessons learned."

I have no doubt that many readers will discover many helpful gems of wisdom within its covers.

Moreover, it is a fun read.

Lama Glenn Mullin,
Seoul, June 2024

Introduction

I am not particularly logical or intelligent. Without guidance from great teachers and less knowledgeable than many, I have always struggled, especially with subjects that required memorization. However, my simplicity prevented arrogance, always seeking lessons in life's events. Gradually, I began observing everything in life just as it is. If I have any talent, it is the "insight" to find order in chaos and the "creativity" to establish new orders.

Over time, I sensed a truth beneath life's phenomena—the magical realization that my life could change through controlling my mind, what people call "creating reality." But creating reality doesn't always happen. Sometimes, despite efforts, nothing changes.

I started exploring why sometimes things happen as if by magic, and other times, no change occurs. I was curious about the order behind our experienced world. There seemed to be an order and moments when my consciousness changed it. Through continuous experiments, I found that the best reality creation occurs when I stay within certain internal boundaries. These questions and experiments were the groundwork for this book.

I experienced miracles when I emptied my mind, maintaining that subtle boundary. After discovering the secret spot of "emptiness," I began practicing the observation of my mind's boundaries. Later, I learned this practice was part of *Vipassana* meditation, quantum physics, and the essence of all philosophy and thoughts. All wisdom points to the same place.

Many desire the ability to create reality, but is not that extraordinary. We are all creating our lives every moment. The difference is whether we are aware of it or not. Often, people are unaware that they are the causes of their own situations.

Most of life's troublesome issues begin with "greed" and are intensified by "bias." Eliminate these two, and problems magically disappear. Desires effortlessly come true when greed and bias are let go, often in unexpected ways. This is the miracle of the "Desireless Desire," the magic of Zero.

Of course, earnestly desiring and working hard could achieve goals. But if you could achieve them without desperately wishing, which would you choose? This realization brings significant changes, transcending all norms and freeing us from the compulsion that we "must do" something.

Can everything fall into place? Can we be free "without doing anything?"

We fear this simple proposition. If we relax, we fear we'll fall behind, unable to catch up. But it's not true. Removing the fundamental fear within us allows us to exist as our "true selves," unlimited and undistorted. This is the existence of a reality creator.

Living effortlessly, my life has become simple. What I need, whether objects, money, or experiences, appears when necessary. I no longer keep a wish list or yearn for things; everything unfolds perfectly. Of course, life has its challenges, but knowing how to overcome them, I am never afraid.

In the eight years after publishing *Zero*, I've created much. Initially, *Zero* was just a book sharing my small experiences. But by creating a culture around singing bowls, I became more certain of the Zero System's existence. For eight years, I lived practicing Zero, solving challenges as they came. The Zero System melded into healing and meditation, creating a vast singing bowl system. Understanding one essence makes applying it elsewhere easy. The Zero System is a solid principle applicable in all life's aspects. This book contains all I've experienced, felt, and learned about creating reality.

What do you truly wish for?

My current answer is, "I want to keep emptying myself to the point of not even knowing I exist." It's not about giving up myself but fully entrusting myself to the universe's flow. To experience the miracle of life's perfection. Our lives are genuinely perfect! Hoping to share these insights in simple language, I began rewriting this book.

Everything in this world starts and ends with me. This is the essence of all philosophies. Therefore, I believe inner peace will solve all of life's problems. I hope more people break free from the limits of their own thoughts, find inner peace, and actively create lives of true happiness.

Explaining how I came to understand the principles of the Zero System is not easy. It wasn't learned from others. One day, inspiration struck, and all laws I knew started connecting. I gratefully accepted this cosmic gift, carefully ensuring it wasn't distorted as I translated it into my language. I became convinced that this was my role to play. Yet, in doing so, I continually emptied myself. Even now, as I write this, I am still emptying myself. I simply hope that the will of the heavens is fully conveyed through "Desireless Desire."

We live in what could be the last era of human life on Earth. Humanity must consciously expand and leap forward. If we continue living with our current lifestyle and level of consciousness, humanity will bring about its own end. We need to abandon our selfish ways and choose a more natural, effortless way of living that promotes coexistence with all. Moving beyond the concept of "I" to an expanded consciousness of "We are one"—this is the "Mind of the Universe." When we become the Mind of the Universe and look at ourselves from a broader perspective, we see things that were invisible when trapped within the "I."

I have named the way the universe operates the "Zero System." By understanding and applying the principles of the Zero System, I hope that everyone realizes their full potential and lives a freer life. I dedicate this book to the complete souls living on this Earth, wishing they awaken to their full possibilities.

May 6, 2020

Chunsia

Invitation to the Zero System

To possess the world and shape it will fail.
The world is a spiritual vessel and cannot be improved.
Those who tamper with it will spoil it; those who grasp at it will lose it.

For things sometimes lead, sometimes follow,
sometimes breathe gently, sometimes strain,
sometimes are strong, sometimes weak,
sometimes destroy, sometimes are destroyed.

Therefore, the sage avoids extremes, excess, and complacency.

— *Tao Te Ching*, Chapter 29

A Letter from the Future

Hi there! Great to meet you.

I'm Chunsia. I've been waiting for you for a long time.

I've come from your future. I always believed you'd read my letter when the time was right, and here we are today! I'm thrilled we've finally met.

Can't remember me? Don't worry about it. I was the same at first. But I'm glad we're reconnecting through this book. Now, let's start crafting a new future together. As time passes, you'll begin to recall more about us.

During our time apart, I discovered the secrets of this world.
Curious about what I found? The thing we've been searching for all along: the secret to achieving everything in life! I found it—the magic lamp exists!

And I figured out why our previous methods didn't work. We have to start from scratch. All this time, the genie of the magic lamp was waiting to grant our wishes, but we just didn't realize it! To properly command the genie, there are three rules to follow:

1. **Have 100% belief in the genie's existence.**
2. **Know the right way to make wishes to the genie.**
3. **Realize that you don't actually need to make wishes to the genie at all.**

Do you understand the first two but feel confused by the third? It's a bit hard to explain ... but it's the key to creation!

Also, remember, there are five reasons why creation might not happen. If even one element is missing, creation can't occur properly:

1. **Rule: Not knowing the universe's laws that trigger magic.**
2. **Wish: Not knowing what your true wish is.**
3. **System: Lacking understanding of the system that moves the universe.**
4. **Order: Not knowing how to properly make your wish.**
5. **Right: Failing to exist as a creator.**

Does what I'm saying seem unclear? In this book I explain why creation hasn't been happening and what needs to change. You've longed to know this secret. You thought you were almost there. But we missed something: it's not about the technique, but becoming the "state of existence" that commands the genie. That's the real secret!

A true creator attracts the universe's energy to create reality Once you reach this state, you'll understand everything you've longed to know.

This mysterious world has always been here, and you and I have always known about it. You just lost your memory for a while and entered this fascinating world.

I've carefully explained everything so you can regain all your memories and join me in this future. My book will help you understand what's been wrong all along and how to fix it. I'll introduce you to the "Zero" state anew. Understanding this isn't something you can do just in your head; you have to experience it. That's why I've included "Practice Assignments" you can do in your daily life to help you shift your consciousness. Don't forget to practice them, okay?

We've always looked outside for answers. But everything has always been right here inside us. We were completely off the mark from the start.

I hope you'll remember who you are again.
Then, you can cross your timeline and join me.

I'll always be waiting for you here.

Goodbye from your eternal angel.

Chunsia

The Act of Creating Reality

People believe life is beyond their control, that they only manage their thoughts, emotions, and actions. We often scramble through life as it unfolds. However, everything around you exists solely for you. Surprising, isn't it?

We are not passive beings dragged along by life. Everyone can create their own reality, exactly as they imagine it! Be it money, appearance, relationships, fame, or wealth—anything at all! Easy to say, but why is it so hard in practice? We ourselves are the most fundamental reason why creation doesn't happen. We are limited by our own "imagination" and "fixed ideas."

Boundaries limit our imagination to what we have experienced and thought. These boundaries are fixed ideas embedded in our subconscious, governing our entire system. Every time you wish for something new, it remains "imagination" and not reality. And as each imagination fails, dissatisfaction with life grows.

With what I've learned, you can actually get everything you imagine. But what if you could create even beyond that imagination?

People often find this idea a bit absurd. But this is the reality of creation! True creation becomes possible when you stop limiting your infinite creativity with limited imagination. Only then can you create a reality beyond your imagination.

This book is filled with ways not just to create what you want but to achieve even more. This isn't just my promise—it's a secret of the vast universe.

People often ask me, "So, Chunsia, what exactly have you created?"
I always respond with a smile, "Everything you see right now."

Creation is the act of bringing something into existence from nothing. It ranges from the very small to the very large, and from things directly related to me to those that seem completely unrelated. People think creation is something special, but it's actually a constant, everyday process. We live in it without realizing we are the creators. Creation is not something extraordinary.

Thinking a thought is creation. - **The creation of thought.**

Setting up a meeting or bumping into someone is creation. - **The creation of action.**

Digesting food and excreting waste is creation. - **The creation of transformation.**

Turning an empty bottle into a beautiful vase is creation. - **The creation of repurposing.**

An argument that makes a friendship awkward is creation. - **The creation of relationships.**

Creation happens every moment, from the abstract to the material, encompassing all aspects of life. Whether you've achieved what you wanted or not, that's creation!

I wish you'd think of creation more lightly. It's not just about performing grand miracles like raising the dead or turning water into wine, but also about recognizing small creations like encouraging friends to order and enjoy pizza together. If I am the agent of change, then all that creation was made by me. Open your mind. When you think of creation as something simple, it can actually happen.

Sounds doable, right?

Creation isn't about meddling in every detail of the process, checking every condition and variable. You're not going to worry about who's cooking the pizza, where the ingredients come from, or how the toppings are prepared, right? I just want pizza, and if it appears, that's all I need!

Creation uses the energy of the universe. It does not follow a set recipe. Even if there were a recipe for creation, nobody would know it. Imagine if God tried to explain it all—your head might explode! We just need to let all the ingredients of creation come from somewhere, and let the universe work.

So don't overthink it. The first step in learning to create reality is to let go of the restrictions you've set in your mind. Let's say it starts raining suddenly and you don't have an umbrella. Is it just bad luck? What if a charming person offers to share their umbrella? Or you get a brilliant business idea while walking in the rain? Open your imagination to how many new possibilities can arise!

Now, does creation feel more accessible? You've already created so much in life and experienced unexpected good fortune, thinking it was just coincidence. There are no coincidences. It's just potential in a latent state briefly showing itself in reality according to the laws of the universe.

But even if the reality you wanted is created, I can't guarantee you'll always feel happy. Ironically, even after achieving the desired reality, we sometimes face new lacks and dissatisfactions. That's the irony of life.

How to Use This Book

This book is a guide to awaken you as a creator of your reality. I've put my utmost effort into detailing everything. I know this topic can be a bit challenging, but I hope you don't give up. I sincerely wish that you achieve the greatest outcomes through this book. As I mentioned earlier, to do this, you need to actively practice and regularly check in with your inner self.

"Awareness," an observer-like stance, is the most important thing to understand this book. Most of us are not used to objectively viewing ourselves. Therefore, we fail to recognize our problems and understand how the process of creation is unfolding. I wish I could be there to guide you through each step, but since I can't, why not take up the challenge yourself? If you have people around who are interested in this topic, it would be great to practice together. Working with others can help you identify and overcome each other's issues, leading to faster transformation.

Here are some tips to help you as you read this book:

1. Set aside your thoughts.
Even if unfamiliar concepts arise, don't rush to judge them. Just consider them new information and keep an open mind. You can always look up related information later when you have time. It might turn out to be quite interesting!

2. Format old information and input new data.
At the end of some paragraphs, I've added **Format and Reprogramming.** These are the sections where you need to format the old information in your mind and input new data. As I said, you need to change first. Repeatedly read these sections to view the world anew. They will surely serve as essential information for creating reality.

3. Concrete understanding through practice and verification.
Verify for yourself whether the content of this book really happens in reality. Notice if the experience of creation is continuous or just a one-time thing. Soon, you'll feel that what I'm saying is true! Try to understand the mechanism of creation more deeply by looking at the connections between what you've learned and experienced. Feel free to record it as a diary if needed. Only what you have experienced can truly become yours.

4. Develop your reality-reading ability.

Important messages about the creation process appear in reality as "indicators." These could be people, objects, or experiences. I call the ability to spot these indicators "reality-reading." Develop this skill by using your intuition to identify indicators. You'll start to see the world from a completely new perspective!

5. Don't get lost in experiences, but embody the state of being.

Following this book will lead to "astonishing success." But don't get intoxicated by temporary joy and lose your center. Maintaining the Zero State is crucial. Keep practicing and proving the results until reality creation becomes a natural part of your daily life. The state of freely utilizing the infinite energy of the universe is indescribable. Keep repeating and proving, and eventually, you'll recognize the feeling. That moment will come when you're entirely in the Zero State. Eventually, letting go of even that feeling will enable you to exist as the true creator of your universe.

Once again, remember, to make this book your own, you need to use intuition, not just thought. The universe's system is inherently a transcendent concept. So, try to feel it through your senses and intuition. Then, you'll understand what I'm talking about.

Good luck!

Chapter 1. Rule

The Magical Universe

Virtue comes from following the Tao and Tao alone.
Tao is elusive and evasive.
Elusive, yet image emerges.
Intangible, yet coalescing into form.
Unseen, and yet containing vitality within.
This essence is so real, within it is truth.
From ancient past to present, the Tao is, undiminished.
And so I observe all origins.
How do I know creation?
Because I see.

— *Tao Te Ching*, Chapter 21

The Field of Creation

"In the universe, there is nothing but atoms and empty space; everything else is mere opinion."
—Democritus, Greek Philosopher

Before we talk about creation, we need to understand how astonishing the world we live in is.

Let's think differently about the concepts of "existence" and "nonexistence."

"Is this stone filled?"
"Yes."
"Is space empty?"
"Of course."

Sorry, but that's not the case.

Your body is the empty one, and space is filled. Let me tell you about the strange truth of this world.

The world we live in is actually empty. Wait a minute. I know, this sounds bizarre. If I pointed at your hand right now and said, "This does not exist!" you'd probably think I've lost my mind.

But listen. I'm talking about the very small world of atoms. Let's get into a bit of science. Atoms are composed of a nucleus and electrons. You probably know this much. If we imagine the size of an atom as a football field, then the nucleus would be as small as a pebble in the center of the field. Meanwhile, the orbiting electrons are like tiny specks of dust. Can you imagine how much empty space there is in an atom? This is the world of particles.

But isn't it strange? If that's true, my hand should appear empty, but it doesn't. I've always been curious about this, and now I know why. Looking at a single atom, it indeed exists in a vast empty space. However, matter is made up of an array of countless atoms. Because these innumerable atoms are overlapping, it appears as if the empty space isn't empty at all. It's like if you look at one net, you can see the empty space, but if thousands of nets are layered, the space is no longer visible. Each atom emits electromagnetic waves that match the frequency of light when it hits an object. This is the unique vibration of atoms. Atoms create interference patterns with each other. The vibrations of these many empty atoms ultimately construct the macroscopic world of my body.

In one of the oldest scriptures in Korea, *Samil Sin Go* 三一神誥, there is a passage that says ...

天訓
Heaven's Teaching

主若曰 咨爾衆 蒼蒼 非天 玄玄 非天
天 無形質 無端倪 無上下四方
虛虛空空 無不在 無不容

The blue expanse is not the sky, nor is the profound void.
The sky has no form or substance, no beginning or end, no up, down, or sides. It is empty, yet there is no place where it is not; it contains everything.

Enlightened ones perhaps already knew that the world we live in is empty. We usually only see what is visible. But the interesting thing is, this emptiness is not really empty. Later on, I'll explain in detail, but the universe that envelops everything is full of immense energy from the Zero Point Field. The space we think is empty is not empty at all. It's just that we are unable to detect this abundant energy.

In the Buddhist scripture Heart Sutra, it is said, "色卽是空 空卽是色 (Form is emptiness, emptiness is form)." It seems to contradict what we see, but modern science says the same.

Have you seen the movie *The Matrix*? Remember when Neo visits the Oracle and meets other potential Ones? Among them, a child practicing spoon-bending tells Neo not to try bending the spoon as it is impossible and says, "There is no spoon." Once Neo understands this puzzling statement, he becomes able to bend the spoon. Creating reality doesn't mean going against established physical laws. It's impossible to go against what has been determined. But if the spoon never existed and everything is just empty space? Then whether the spoon is bent or melted doesn't matter.

Everything starts from "this world does not exist."
We live in an incredible world. We must not forget that this world, which appears empty, is the real world. Then you can be like Neo!

Format and Reprogramming

Format
1. Everything is fixed → X
2. What is visible is all there is → X
3. Phenomena are unchanging → X

Reprogramming
1. All matter is empty.
2. The world I perceive does not exist.
3. If you flip everything in the world upside down, the invisible world becomes visible.

The Law of Attraction

"If you really want something, the whole universe conspires in helping you to achieve it."
—Paulo Coelho, *The Alchemist*

How far do you agree with the statement, "Life unfolds according to my thoughts?" Perhaps you agree in some cases, but think other things are absolutely impossible. Surprisingly, this world can indeed unfold as we think.

Now, let's talk about the unseen aspects of the world we live in.

Imagine you're walking down the street and you suddenly encounter the person of your dreams. Instantly, your heart starts racing and an instinctive feeling spreads throughout your body. You quickly turn your head to see them again, but they're nowhere to be found. Where did they go? From that moment, your mind is consumed with thoughts of them.

When we perceive something, various parts of our brain generate different brainwaves. Brainwaves are electrical oscillations that occur due to brain activity. Our brainwaves speed up when we're tense or deep in thought, and slow down when we're relaxed. Just like how your mind was filled with thoughts of her after she appeared, and you might spend the whole day thinking about her.

Whenever we think, numerous signals begin to emerge in our brain. The brainwaves start to dance. But think of these oscillations as a form of energy. Every thought we have emits its own unique wave into the universe. Like, "Ah, she was so beautiful!"

Our usual thoughts might not seem like a significant energy. You'll soon forget her, after all. However, occasionally we create intense thought energies. For instance, when you think of someone you truly despise, like your sworn enemy ... who never fades from your memory, even over time.

Intense thoughts that are repeated over time create isolated wave patterns, called solitons. Solitons are like waves that maintain their form without being affected by other waves, almost like particles. They possess tremendous power, enough to influence the physical world. While regular waves may cancel out or change upon meeting others, these intense waves are so powerful, they seem like living energy. Think of haunted souls with deep grudges in folklore. Such intense thoughts can exist as physical energy entities. I believe these intense solitons can become mental energies powerful enough to affect the material world. Whether you believe in ghosts or not, the power of deep grudges can be terrifying!

Such repeated thoughts create solitons. These are often referred to as desires. Prominent energy solitons include prayers, hymns, mantras, and spells, practiced in religions. When countless people over many years share the same thoughts and produce the same sounds, a powerful energy is formed. Even now, people around the world are praying, chanting, and reciting spells.

But more important than these created solitons is the concept of resonance. Vibrations resonate with similar vibrations, amplifying into significant energy. This can be described as "attraction." You're likely drawn to people who share similarities with you. Your surroundings are probably filled with people who you think are compatible with you. Those who don't fit have already left your circle. These are examples of the attraction that occurs around us. But when like-minded people come together, they can achieve greater things than when alone.

Ever heard of the Butterfly Effect in chaos theory? A butterfly's wing flap in Beijing could cause a hurricane in the US! That small wing flap can resonate with others, creating immense energy. That's the real power of resonance.

If you chant a mantra, for instance, you momentarily resonate with an energy field already created by many before you. Thanks to that, you can harness a part of that energy. Prayers being answered or miracles occurring can be due to the utilization of such energy. Our thoughts and vibrations transcend space and time, instantly transmitting everywhere.

Similarly, what if a person's faint thought is repeatedly expressed in the same imagination and words over a long period? That vibration will start to materialize into universal energy. The key is persistence. It doesn't happen just once. The phrase "dreams come true if you wish hard enough" fits here. Intense repetition over a long time creates solitons. Once a soliton is formed, it resonates to attract something similar in reality. It's like selecting a TV channel. If I want to watch sports, I tune into that broadcast frequency. Tuning in means "synchronizing frequencies." It's about creating resonance. When the weak energy of a signal resonates with the energy of a receiver, the faint information amplifies into a clear image on the screen. In the same way, we experience reality that resonates with our consciousness: meeting a new person similar to our thoughts, or encountering someone who can realize our dreams. In reality, new opportunities resonate with our consciousness, creating new manifestations. Perhaps your current reality is just a reflection being created by resonating with your conscious energy.

Every thought you have is crucial.

Imagine two people: one who always says, “It will be okay,” and another who constantly says, “I feel like I’m going to die.” How different do you think their futures will be? Speech is an expression of a person’s beliefs as vibrations. A person with a consistently positive belief system and another with a negative one will live very different lives, won’t they? They express their beliefs through their words, and consequently through their actions. Eventually, they will encounter a destiny that aligns with their beliefs, words, and actions.

This is the essence of creating reality through the “Law of Attraction”!

Format and Reprogramming

Format:
1. Thoughts are not significant → X
2. Can thoughts really manifest into reality? → X

Reprogramming
1. Every thought generates its unique wave pattern.
2. These waves resonate and amplify when they encounter similar waves.
3. Repeated, intense thoughts create actual power.

The Strange World of the Micro

"What we observe is not nature itself,
but nature exposed to our method of questioning."
—Theoretical physicist Werner Heisenberg

Modern physics consistently strives to find the smallest units that make up matter. They've split atoms further and further until they discovered quarks, which turned out to be ambiguous, sometimes acting like particles, sometimes like waves. These enigmatic entities are known as quanta. Quanta made scientists, who prefer clear-cut answers, quite uncomfortable.

Now, let me tell you an even more fascinating story. There was a legendary experiment in quantum physics, hailed as one of the greatest discoveries of the twentieth century, known as the double-slit experiment. Physicists were observing photons passing through two slits in a wall and hitting a screen behind

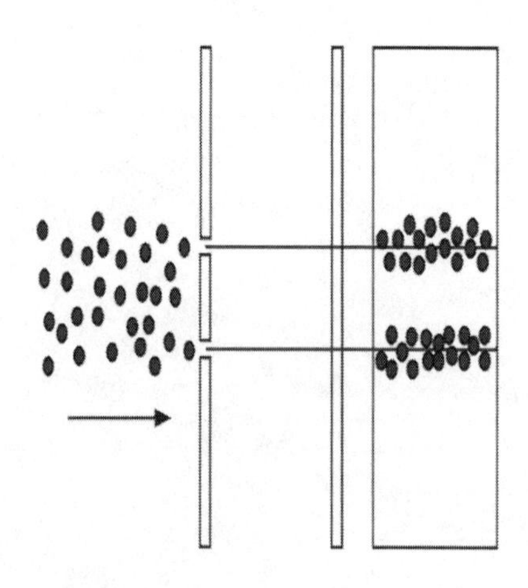

Pattern created by particles passing through a double-slit

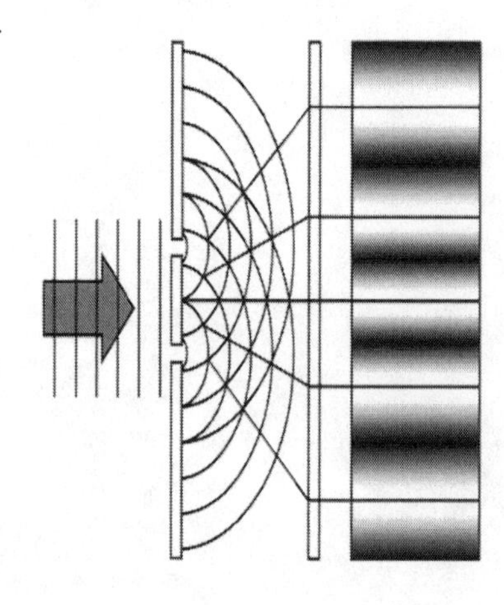

Pattern created by waves passing through a double-slit

it. They were trying to determine what kind of mark they would leave. If light were a particle, it would create two lines on the screen, mimicking the slits. If

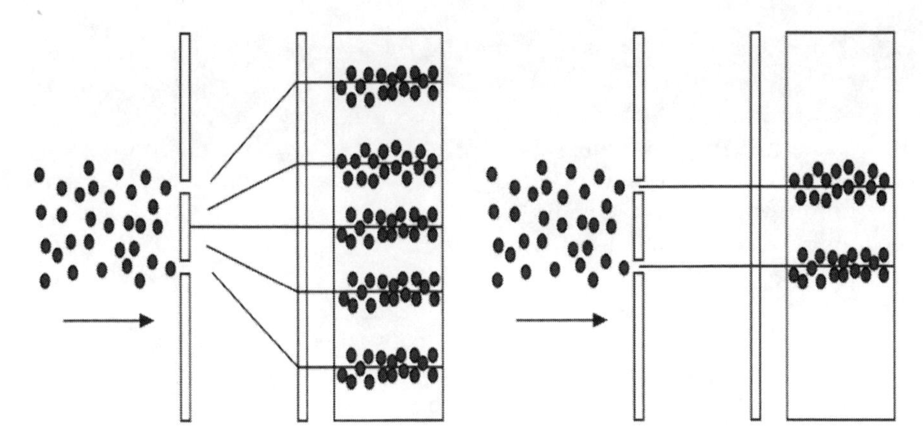

Electrons passing through a double-slit create different patterns depending on the observer's intent.

it were a wave, a more complex interference pattern would emerge.

To observe more precisely, they placed photon detectors next to the slits. The results were intriguing. When observed, the photons acted like particles, creating two lines, and when not observed, their wave nature reappeared, creating interference patterns. Photons can pass through multiple places simultaneously and overlap, but the moment they sensed they were being observed, they changed their stance and became fixed like particles. Not just photons but also electrons, positrons, neutrons, quarks, etc. all exhibited particle-wave duality, changing their nature moment by moment. All these micro-world entities showed that observation affects the state of quanta, a phenomenon called the "observer effect."

I want to share with you a poem, "Flower" by Kim Chun-soo.

Before I called his name,
he was nothing
more than a gesture.

When I called his name,
he came to me
and became a flower.

The German physicist Werner Heisenberg imagined a unique experiment: observing an electron with a highly magnifying microscope. Normally, visible light can't observe an electron, so he theorized using short-wavelength gamma rays. He discovered that when a tiny particle like an electron collides with gamma rays, its direction and momentum changed. Even with the most precise tools, it was impossible to measure both the velocity and position of a particle simultaneously. Additionally, particles themselves don't always have a defined speed and position.

Quantum physics, the study of the laws of the micro-world, is relatively new, with only about 120 years of history. Despite this, quantum physicists have tirelessly worked to understand the principles of this tiny dimension to unravel the greatest secrets of the universe. There are many hypotheses in quantum physics, **but the most widely accepted view is that in the micro-world, particles exhibit wave-particle duality, are not fixed, and are difficult to observe accurately with current technology.** In other words, no one can definitively say what they are. It's the pinnacle of ambiguity.

Physicist Niels Bohr argued that quanta follow the principle of complementarity. Particles and waves are complementary; they exist as one entity. When a particle's nature is observed (or fixed as a particle), its wave nature disappears (or becomes unmeasurable), and vice versa. Until observed, even this isn't determined. Quantum theory suggests that "the only thing that exists is the observed particle." **Before our observation, everything exists in a state of potential. From the moment we observe and through our observation, our consciousness instantly creates reality. Isn't it exciting? It means that anything is possible until it's defined.**

Format and Reprogramming

Format
1. Objects cannot change. → X

Reprogramming
1. Everything possesses both particle and wave properties.
2. Nothing is definite until it is observed.
3. Everything exists in a state of infinite possibilities.

An Unusual Middle School Student's Test Scores

I'll tell you about when I first became aware of the principle that reality changes according to my intention. This peculiar experiment was the starting point for me discovering this law, so it's a crucial clue for you to understand what I'm saying.

As a child, I believed Christianity. Back then, I firmly believed that God loved me so much that He would grant everything I desired. It's a common experience, right? Whenever I wanted something, I always prayed to God. Just like everyone else, I prayed for trivial things when I needed them. And I always got what I wanted, whether it was money, rubber bands, or even plastic bags—all the small things. I was a pure child with the strong backing of a God who would grant me whatever I desired. However, as I grew older, I became disillusioned with Christianity due to its materialistic practices. I began to think that maybe God wasn't a personified being like Jesus, but a broader and more expansive universe.

I attended a girls' middle school. I was quite introverted and loved comics. I enjoyed sitting quietly at my desk, doodling. I was a kid with my own unique world. My appearance? Nothing special. And my studies? I did fairly well in middle school, primarily because I had nothing else to do besides drawing. I was just a good girl who listened to my mother.

It was probably around the time of my midterms. Naturally, there was pressure about the exams, and I desperately wanted to do well. I wanted to make my mother proud with perfect scores. Back then, school rankings were publicly displayed, and grades were marked as "Excellent," "Good," "Average," "Poor," or "Fail." Getting a "Good" on the report card was a blow to my pride.

But despite knowing all the answers, I always made a few mistakes, ending up with "Average" scores. I could have gotten everything right if I had been more careful. At that time, I was quite obsessed with getting perfect scores. It felt like such a waste to get a "Good" for missing out on questions I knew.

Whenever I got the test and thought, "This is easy. I can ace this," I always scored "Average." It was my complacency. Conversely, when I found the test difficult, thinking, "Wow, this is tough," I scored better than expected, even "Good" despite not knowing some answers.

I started to have a strange suspicion about these results. So, to improve my grades, I began analyzing what the problem was and its cause. And I came up with this hypothesis.

Thinking "it's easy" → Overconfidence (I'll get Excellent!) → Carelessness → Making mistakes (Average)
Thinking "it's difficult" → Not overconfident (I'll only get Average...) → Becoming cautious → Achieving better scores (Good or higher)

I didn't just think, "I made mistakes, I need to be more careful." Instead, I believed that my mindset influenced the outcome of my exams.

Therefore, I decided not to prematurely determine the results as soon as I received the exam. I would not judge the questions. I would approach each as if it were new or unknown. I believed that expecting an outcome in advance and having a preconceived notion was what led to mistakes. Although I knew the answers, I made errors, right?

Moreover, seeing that the subjects where I was confident of getting "Excellent" often scored lower, I thought that nothing could be certain until the results came out. So, I decided to think this way:

"Maybe my preconceived mindset and expectations altered the exam results. Until the results are announced, nothing is set in stone. It could be higher or lower. Until I verify it, nobody knows. So, let's drop the expectation that I got everything right and keep an open mind. If my preconceived notions are indeed altering reality, then I won't have any until the exam results are out, and I'll approach the exams with a serious attitude."

After that, every time I took a test, I started to control my mind.

"I don't know anything. I can't predict what will be on this test. I'll just do my best to solve each problem as if I'm seeing it for the first time. And I won't assert anything about the outcome. The results are unknown to everyone until they are published. If I keep an open mind, I will likely get much better scores than expected. I will keep all possibilities open."

Do you know what happened? The scores actually improved. Subjects where I let go of my expectations resulted in higher scores. I even started getting the "Excellent" I wanted. This led me to conclude:

- Reality isn't predetermined until it happens.
- Reality reflects my mind, influenced by my subconscious beliefs.

These were my own magical rules. I applied these two rules to view life differently, focusing more on being right and open-minded. And as I believed and observed, reality began to change accordingly!

It might sound like a strange, almost forced rationalization of my exam performance. But remember what I told you earlier? According to quantum physicists, in the world of particles, nothing is decided until observed. Let's apply this to our reality. Until an event is confirmed by me, no one knows the truth. Exam scores could be "Fail," "Excellent," or "Average." Even after I submitted my paper, the score wasn't set until I observed it.

This leads to a question:
If the result can change infinitely even after submission, what determines it?
Your skills? That's too one-dimensional.
I thought it was "consciousness."

So, I began to observe my consciousness.

For over twenty years, I've viewed the world through these "rules," discovering that reality indeed dances to the tune of my mind. You might wonder:

Am I experiencing this reality because I set these rules?
Or, are these rules a universal order for everyone?

To find the answer, one must first have an open mind.
The world is like quantum particles, nothing is definitively determined.

Format and Reprogramming

Format

1. Every event is predetermined by my actions. → X

Reprogramming

1. The future is not determined until it occurs.
2. How I perceive and set rules shapes how the world unfolds.
3. The future can unfold beyond my imagination.

Cinderella's Magic

I have an intriguing fact to share with you. Our bodies can change according to our thoughts. I've successfully experienced this myself. As my confidence in creating reality grew, my appearance began to transform.

Seventeen years ago, when I was nineteen, something magical happened to me. During my middle and high school years, I was just an average girl who didn't stand out. My lack of confidence and unremarkable appearance played a part, and I often consoled myself by thinking, "True beauty is about having a kind heart, not just a pretty face." I wasn't popular with boys and was more like an *otaku*, absorbed in manga.

Girls who don't get attention often dream of becoming beautiful and admired. I was no different, dreaming of becoming an attractive woman someday.

Then, as I gradually strengthened my belief in my power to create reality, my wish to become beautiful started manifesting. It wasn't just a fervent wish but a subtle desire. I knew the rules well and how to apply them. Surprisingly, my appearance changed rapidly within a year. I went from being unnoticed to attracting attention everywhere. Without any significant makeup skills or plastic surgery, I transformed into a shining beauty. I even started modeling. My life took a completely different turn as I became a woman who sparkles wherever she goes.

People often commented, "You look Japanese, like a manga heroine."

Maybe, after immersing myself in manga daily, my dream of living a manga heroine's life became reality.

Even after seventeen years I still maintain my youthful looks.

How could this happen? Everyone experiences changes in their appearance over time. Our face and overall appearance are constantly evolving, responding to the energy changes within us. There were days when I looked older or unattractive, and days when I looked innocent and radiant, especially after meditation or feeling joyful. I realized through these daily changes that a person's appearance is a reflection of their inner energy.

According to Dr. Serge Kahili King, who studied Hawaiian shamanic healing, the Hawaiian word for body, *kino*, means "a clump of highly energized thoughts." He suggests that our bodies, being materialized forms of our deep subconscious intentions, can change as we alter our perceptions, attitudes, and habits. Our bodies are essentially formed by vibrating energy.

Our physical state easily shifts with our thoughts and feelings. When we hold something beneficial in our hand, our subconscious mind increases our body's strength. Conversely, holding something detrimental weakens us. This principle is used in the O-Ring test, which demonstrates what is good for us.

To do the O-Ring test, place the item in one hand. With your other hand, make an O shape by joining your thumb and index finger, pressing the tips together firmly. Then, have someone try to separate your O fingers. If your fingers remain firmly together, that item is beneficial for you. If your fingers easily part, it's detrimental.

Our body and subconscious are closely linked, and our subconscious already knows what is beneficial or harmful for our body, beyond the limitations of our current consciousness. To explain the O-Ring test principle in more detail:

1. Current consciousness recognizes the content of the test: "I am now testing whether this food is suitable for my body. If it's suitable, my fingers will remain strong and not separate easily. If not, they will weaken and separate easily."

2. The content recognized by the current consciousness is conveyed to the individual's subconscious.

3. This is then transmitted to the collective subconscious.

4. The collective subconscious's response or answer is transmitted to the body's energy field.

5. The message from the collective subconscious manifests as the strength of the O-Ring.

We often experience physical discomfort like stomach aches or unexplained headaches just by thinking about an unpleasant boss at work. Doctors refer to these as psychosomatic illnesses. These are conditions where physical symptoms repeatedly occur due to mental stress, despite no apparent physical cause. Conversely, there are cases where patients diagnosed with terminal cancer miraculously recover due to a strong will to live.

The reason our thoughts can have such tangible effects on every corner of our body is due to what is called the "biological matrix." Previously, medical understanding assumed that our body only recognizes information transmitted through the nervous system. However, recent research has revealed that our body's biological matrix allows all information within the body to be exchanged in the form of electrical signals.

Dr. Albert Szent-Gyorgyi, who won the Nobel Prize for the discovery of the biological matrix, announced that this matrix exists within the connective tissue. Connective tissue binds tissues and organs together, encompassing skin, the musculoskeletal system, the nervous system, and internal organs. Think of it as "connected cells" for a clearer understanding.

The connective tissue's biological matrix instantly converts changes in one part of the body into energy, transmitting this information throughout the body via fascial tissue. It's a communication network. Every moment, our brain receives a tremendous amount of information from all over the body through this delicate biological matrix and interprets it. Previously, we thought only the brain controls information, but it turns out that the entire body is a vast network of interconnected information.

Remember the 2009 blockbuster movie *Avatar*? On Pandora, the sacred Home Tree connected all living creatures of the planet, ensuring their protection. Dr. Grace, the botanist, argued that because all plants on Pandora were interconnected through their roots, sharing information, harming the Home Tree could have catastrophic consequences. On Pandora, not just plants but also animals shared thoughts and emotions by physically connecting. It's an effective symbol for the biological matrix.

Biochemist Glen Rein, in his book *Quantum Biology,* states that since the human body is made up of quanta, it follows the principle of particle-wave complementarity. Our body consists of a delicate network of cells and tissues that transmit information. Each cell is an independent entity with its unique energy, and these minuscule electrical signals carrying specific information can induce physical changes. According to German biophysicist Fritz-Albert Popp, cells and tissues are interconnected by a biophoton field—a light energy field emitted by cells, primarily from DNA, called the DNA information-energy field. This interconnected network allows for rapid information exchange.

Valerie Hunt, a professor of rehabilitation medicine at UCLA, argues that information about all parts of the body can be obtained through electrocardiograms or electromyograms, not just the area being tested. This is possible because the information-energy fields of each organ and tissue in the body are interconnected. Understanding each vibration's characteristics allows us to understand the condition of organs. Technologies like InBody or electrocardiogram tests, commonly used in medical practice, are based on these principles, enabling easy monitoring of various bodily conditions.

Consider our body as a massive information system, each part having its unique vibration. If our intentions can influence the quantum world, then changing our bodies with the power of consciousness might be possible. Of course, simply imagining a particular celebrity won't transform your body to match theirs. A major reason for this is the belief in its impossibility. While such a transformation might seem impossible within the confines of a limited mindset, opening up to the possibility can change the narrative.

Embrace the wonder of this amazing world with a pure belief. You might be living in a far more incredible world than you ever imagined.

After realizing that I could control my body and the unfolding scenarios in front of me using mental energy, I truly began to enjoy the scenarios I created. We have the power to change our reality and even our physical selves through conscious control. This mental strength isn't about forceful concentration but involves a delicate balance, almost like walking a tightrope of energy. I'll explain more about this later.

Format and Reprogramming

Format
1. Accepting the given environment as an unchangeable reality. → X
2. The body and mind are separate entities. → X

Reprogramming
1. The body can be controlled by the mind.
2. The environment I am in can also be controlled by the mind.
3. Everything is organically connected.
4. Everything that exists in this world is energy.

The Path Forms as I Walk

I've lived my life doing what I wanted, rather than following a path set by others, possibly because I understood how to create my reality.

As a child, I was like a character in a comic book, always dreaming of a fantastical life. I enjoyed imagining and dreaming. In middle school, I did well academically, but I preferred drawing. When it was time for high school, I wanted to attend a vocational school to study art and design. However, my teachers and parents, seeing my academic potential, urged me to pursue further studies and attend a better university. I was unconcerned about societal views and wanted to start my desired path early, but in the end, not wanting to hurt my parents, I enrolled in a regular high school. As my interest in my studies waned, I drifted and focused on drawing. My grades declined.

My senior year, I made a choice that was true to myself: I joined a vocational class, which involved attending another school for a year to acquire a specific skill. While my peers were focused on university entrance exams, I was engrossed in drawing, learning about video production and animation, and creating two short animations. I won fifteen awards in national competitions within that year, establishing myself as a promising talent. I relished the freedom to fully engage in what I wanted to do, though my parents were increasingly worried about my future. To reassure them, I promised to fulfill my mother's wish and get into a university.

Despite limited study time, I kept my promise. I studied for the university entrance exams on my own and managed to get into Kyung Hee University's Multimedia Department. My unorthodox path made me a bit of an oddity in both the vocational and regular schools. I've always been the type to forge ahead on my chosen path, embracing life as a game, without any fears or "shoulds."

In my journey from a high school senior to a university student, I was filled with immense confidence and energy, feeling all the energy being drawn towards me. My appearance also improved significantly. As I achieved my goals, I felt I had a special power of creation.

After enrolling in university, I aspired to become the youngest video director. I created numerous videos and gained recognition in various fields. I could have continued successfully in this field, but at twenty-two, I decided to study

English in India for six months. This decision marked a turning point in my life, leading me to delve into the spiritual world. I became immersed in healing and meditation, abandoning my previous path to embark on a new adventure. This early realization led me to start my Zen Healing Shop. At that time, Korea lacked interest in healing and meditation. When I joined Kyung Hee University's Graduate School of Alternative Medicine in 2009, many questioned my decision, as I once again ventured on a path less traveled.

My choices have rarely been supported by others; it felt like walking alone in an uncharted field. But it wasn't too difficult, thanks to my self-support and a certain conviction. Despite everyone's disapproval, I quietly embarked on another adventure, changing my life.

After a decade in healing, I became a recognized expert. Initially, people didn't understand singing bowls. Meditation was seen as sitting in silence, making singing bowls seem unnecessary. However, I worked tirelessly to introduce singing bowl meditation as an accessible practice. My innovative lying-down meditation method caught on because many found traditional meditation challenging. I broke the mold of conventional meditation and promoted my method, the "Chunsia-style Meditation," making meditation easy and enjoyable for everyone. My efforts led to the widespread adoption of singing bowls in Korean corporations, creating a new meditation genre. I established the entire singing bowl system, and many began to follow the path I created.

However, this is just the beginning of my creation journey. I know that wherever I walk, new paths emerge, and others follow, turning these paths into real roads. This is the essence of creation.

I've never seen "limits," only "goals." I simply walked through the unfolding journey without any fear. My approach was simple:

"I want to do this." → "Okay, just do it."

I acted on my impulses without being swayed by external perceptions, firmly believing in myself and my power to create. To build your world, you must be your life's designer. You are the protagonist of your universe, endowed with immense latent power. To harness this power, you need a clear goal. I had distinct goals and channeled all my energy into creating my world, never doubting and enjoying every manifestation.

When the belief that your life unfolds as you wish becomes a reality, that belief starts to work. It's the difference between being the "designer of your life" and an "extra believing in the designer."

Now, here's what you should do:

Format and Reprogramming

Format

1. The world has a predefined standard of living well → X

Reprogramming

1. Every choice I make is right.
2. I am the creator of my own life.
3. Every choice I make has its own value.

Life Is a Giant Holograph

Life, to me, is like a game. Our world can be likened to a holographic universe, created and manifested through laser light technology, capturing three-dimensional images without a lens. How did I come to know this? By continuously observing how my consciousness and emotions reflect and shape my life. This observation has been an experiment, a practice of manipulating subtle energies.

Let me share some of my experiences.

When I first delved into the spiritual world, I hopped between various meditation groups until I encountered Tao. A friend introduced me to this practice, which piqued my curiosity. I eventually began practicing it seriously. The group I joined was a pseudo-religious cult, which had a unique philosophy:

We are living in the end times, where various calamities, especially untreatable diseases, will strike. Only those who practice "Special Healing Power" (attaining medical power through spiritual practice) can save people from these plagues. To gain Special Healing Power, one must diligently practice the Tao.

Being deeply interested in healing, I was fascinated by the concept of Special Healing Power, a special ability to save lives. It seemed like a noble pursuit, even more impressive than being a doctor in an era of incurable diseases. So, I devoted myself to practicing The Tao. The practice in this group, which involved chanting Mantra and spreading the teachings (*Podeok*: a form of proselytization), seemed unique. In hindsight, it was more about expanding the group, but I was indoctrinated to believe it was the only path to Special Healing Power.

After a year of dedicated practice, I had some success within the group. However, I eventually left upon realizing the contradictions in the teachings and the cult-like behaviors. Still, I learned one important lesson:

Our reality is a reflection of our mind.

Though I always believed my mind could influence reality, this intensive year of practice made me experience it more profoundly. The practice, though bizarre, was a form of spiritual discipline. Facing rejection and scorn from strangers, learning to humble oneself—it's all part of "lowering the heart" in spiritual practice. It was a unique experience I wouldn't have otherwise had.

Later, I began to view reality more experientially, erasing the boundaries of good and bad. Even a pseudo-religious experience had its lessons. It taught me that experiences are just that—experiences. How I perceive them is what truly matters. "Even if it's a negative experience, it's just a "happening," not inherently bad," became my light-hearted approach.

Looking back, I've lived a relatively smooth and comfortable life, more illuminated moments than dark. I've rarely faced failure because, as I said, I've mostly achieved what I wanted.

Someday someone told me:

"Chunsia, you are like a flower in a greenhouse. You haven't seen all aspects of life, so you can't fully comprehend true healing. Do people come to you for your skills or for your looks?"

This statement was a shock. It resonated with me to some extent. Despite my early spiritual insights and counseling many from a young age, I lacked a complete understanding of the suffering of middle-aged people. And at that time, my classes were mainly attended by men attracted to my appearance.

I decided to further develop my skills, to be recognized for my abilities, not my looks. Like Siddhartha Gautama, who left his royal life to understand the world, I chose to experience life's darkness. I deliberately sought experiences that most would find bad or degrading, wanting to transform from a delicate greenhouse flower into a resilient wildflower.

I chose to be a tragic heroine and experimented with smoking, clubbing, dating married men, working in bars, giving up things I cherished, living like a pauper ...

These were voluntary experiences, not excuses. I wanted to feel a range of emotions and break my own stereotypes. Although unpleasant, they were meaningful spiritual exercises, allowing me to experience the other half of life I didn't know. After these experiences, **I understood that life has no good or evil, wealth or poverty—these are just concepts we create. I realized I could play any role in life, everything was just an experience.**

When bad things happened, I didn't perceive them as negative but just as experiences. This freed me from emotional residues, making my past memories feel like illusions. I felt liberated from everything.

Then came a significant realization, leading to the main content of this book, the concept of "Zero." By experiencing duality, I fully understood both extremes, which made me stronger and wiser. I learned to make choices in life and to accept what has already happened without fear. I truly grasped the meaning of creation.

Everything existed within my mind.

The Brain Evolves Infinitely

Our thoughts and actions are constantly under the control of the brain, influenced by hormones. Emotions like laughter and anger are simply the results of hormones like endorphins and adrenaline. In essence, "human life occurs within the brain" isn't an overstatement.

The influential Spanish neuroanatomist Santiago Ramón y Cajal once declared, "The adult brain's neural circuits are fixed and cannot change." This view was widely accepted in neuroscience, leading to the belief that human brains are predetermined by genetics at birth and deteriorate with age. However, modern neuroscience and neurophysiology have revealed the brain's continuous developmental potential, known as neuroplasticity. The brain, maintaining itself through cell division, can change its function and structure in response to external stimuli and environmental changes.

A notable example is the Silver Spring monkeys experiment by Dr. Edward Taub in the 1970s. He conducted a study severing the sensory nerves of monkeys' arms to measure sensory loss. At the time, scientists believed that if sensory nerves were cut, the corresponding brain functions would cease. Although initially suspended due to animal cruelty allegations, an opportunity arose a decade later to dissect the brain of Billy, a monkey who had undergone sensory nerve severance in both arms. Astonishingly, active electrical signals were found in the brain areas that perceived arm sensations, but these were in response to facial stimuli, not arm sensations. The brain had adapted its function to new stimuli, indicating its ability to change and adapt—a natural self-healing property.

This experiment suggested that the brain constantly evolves to overcome structural flaws. Billy, along with other monkeys like Augustus, Domitian, and Big Boy, exhibited this adaptability. The brain adapts and changes, functioning like a perfect, self-sustaining organic system—a miniature universe in itself.

Our brain is a source of unlimited potential and creativity, not only a physical hardware but also the seat of our consciousness. If the heart is the power source of our body, the brain is that of our consciousness. Consciousness, the observer within us, is crucial because it's the driving force behind physical changes in our environment.

Understanding consciousness can be challenging. Think of it as the "observing subject." When you wake up, your body transitions from an unconscious to a conscious state, powering up. Consciousness is not only the eye that sees the world but also the one that observes ourselves—our thoughts, feelings, intentions, actions, and emotions.

Sometimes, consciousness is accompanied by a mind. When you desire something, whose mind is it? Is it "Chunsia's" mind, or the mind of the "consciousness observing Chunsia?"

Now, let's talk about what you've known as "you" and the true nature of your original consciousness. To influence the world, consciousness must first command. It's the entity that initiates changes in the quantum realm.

Neuroscientist Dr. Fred Gage stated, "Our brain changes with the environment and experiences. Your current self is shaped by your environment and accumulated experiences." If the past me has created the present me, then creating a new future in this moment is entirely possible. You can choose what to create now. Just like brain plasticity, we have the power to change even the functions of our brain.

So why doesn't everything go as planned in such a marvelous world?
It's because you're not making those choices.

Chapter 2. Wish

What Is My True Desire?

Colors blind the eye.
Tones deafen the ear.
Flavors deaden the taste.
Pursuing maddens the mind.
Desires wither the heart

The sage follows intuition, not sensation.
Ignore the latter, heed the former.

— *Tao Te Ching*, Chapter 12

Why Visualizations Don't Materialize

The main theme of *The Secret* is the Law of Attraction, which means "if you desire something fervently, it will come true." You might already be aware of this. So has your life changed?

But, you know what's interesting? **Our brain cannot differentiate between illusion and reality. It perceives deeply ingrained images in our imagination as actual facts.** If you repeatedly imagine "I am beautiful," your brain will start recognizing you as a truly beautiful person. It seems our brain might be optimized to perceive the holography of this world. Visualization techniques are based on this principle and are widely used in sports, art, hypnosis, and self-improvement to enhance skills. I've mentioned this before, haven't I? When we think intensely, that energy of thought becomes real. The brain, while recognizing our thoughts and imaginations as simple facts, also affects the micro-world and situations. This is an undeniable fact.

However, the effect does not manifest equally for everyone. The reason varies from person to person depending on the energy of hindrance they emit. For someone with high hindering energy, the effect might be minimal, and for those with less, the effect could be more significant.

Just a simple "want" is not enough for creating reality. We all have different belief systems, and these differences affect our ability to create reality. Some people are positive, while others are negative; some are lazy, others are passionate. These variables significantly influence people's creative power.

Our brain accepts images in our imagination as real, but even in this process, there are differences based on situations. We usually look for evidence before accepting something as a fact, and these evidences accumulate through experiences. Therefore, an individual's experiences play a crucial role in shaping their reality.

Even if you try to keep a positive mindset, thinking, "I should live positively!" or, "What I think will come true!," doubts and distrust arising from your experiences might surface, leading you to think, "Will it really happen?" or, "It's useless. Such things can't easily happen." These thoughts prevent your intense visualization and desires from becoming a reality.

Our conscious and subconscious commands are constantly conflicting. Most of the decisions we believe we make consciously are, in fact, influenced by our subconscious. It's impossible to make decisions that go beyond our subconscious. That's why understanding and controlling the subconscious is essential.

You should reflect on your life journey and thinking patterns to understand the structure of your consciousness. If negative thoughts and emotions continue to surface from deep within, hindering the actualization of your reality, no matter how much you fervently wish and concentrate, it will be difficult to fully realize those thoughts.

Also, even if you fervently wish for something, if that desire is distorted, the universal energy will not move according to your will. A distorted desire means what you think you want might not be what you *truly* desire. In essence, it's a wrong command.

Now, I will tell you about the universal system of human consciousness. You need to know how your consciousness system is structured to execute the commands for what you truly want.

Deficiency Energy vs. Desire Energy

Thoughts don't always attract only good things; they can attract bad things too. Sometimes, the things we worry about seem to happen more frequently than the things we hope for. Why is that?

When we're consumed by worry, a chain of negative thoughts can seem endless. We might become unnecessarily anxious or suspicious, leading to misunderstandings and mishaps, even when nothing bad has happened. And when something bad does happen, we quickly rationalize it with, "I knew this would happen ... I had a bad feeling!" This is a way of justifying and avoiding responsibility. But was the bad outcome really a matter of fate that we intuited, or did we create it ourselves through excessive worry? It's likely the latter. The power of human consciousness in creating reality is quite remarkable. In this respect, everyone might be a magician!

Desiring something implies a need for it, which in turn suggests its absence. Let's take an example. When we need money, we wish to have more of it. However, the more we think about money, the more we grow conscious of our "lack."

Nobody wishes to acquire something they already have. Every wish stems from the thought that "my current reality is insufficient and incomplete."

So, the more intensely we desire something, the more we inadvertently generate the energy of "lack" in our subconscious. This is why, despite attracting what we wish for, real change often does not occur in our lives. It's because we continue to create this hindering energy. Even though we consciously wish for a lot of money, subconsciously, the more we desire, the more we end up thinking about "the lack of money" instead of "having money."

We can never achieve our desires if we're only creating negative energy of lack. It's foolish to reinforce the opposite energy while trying to create reality. To break this chronic pattern and achieve our desires, we need to eliminate the energy of lack and amplify the energy of desire.

The universe fulfills whatever we wish for, without any moral judgment. Only the quantity of energy matters. If you have a deep and genuine desire, you need to increase the energy of desire instead of the energy of lack. Then, the universe will surely fulfill your wish!

If you have a long-held dream that still feels distant, it's worth looking inward. Perhaps you're unknowingly creating a tremendous amount of "lack" energy.

The Structure of the Mind

The unconscious, which we cannot perceive, is like a massive iceberg hidden beneath the water, and the conscious mind that we can perceive is just the tip of the iceberg visible above the surface.
—Sigmund Freud

Our unconscious is filled with hidden information. It wouldn't be an exaggeration to say that most of what we call consciousness is actually unconscious. Freud wanted to explore the human unconscious through free association. Free association is where you lie down comfortably and mutter freely any words or thoughts that come to mind. When you release the controls and defenses of consciousness, the information from the unconscious begins to pop into current consciousness. That's the moment when you encounter the unconscious within. Since a significant part of our chosen language and actions originates from the unconscious, it's possible to explore the unconscious by observing our current consciousness in reverse.

Current consciousness may celebrate a success, but the iceberg underwater is etched with experiences leading up to the moment of victory, such as struggles without money, painful times of hard practice, the agony of giving up things you love, frustrations, and pain. All the experiences accumulated at every moment of life form the large body of the unconscious.

Even if it's so long ago that you can't remember, all experiences are frozen and stored deep within the subconscious. The unconscious constantly influences the current consciousness without even realizing it, under the guise of "lack."

Below is a workbook to awaken the creative power lying dormant within you. I hope that as you fill it out, you will be able to reflect on yourself.

workbook

What Exists in My Unconscious?

1. Write down the ten most positive traits and characteristics you possess.
   Example: I am honest, patient, loving, etc.

2. Now, write down ten traits and characteristics that are the opposite of what you have written.
   Example: Impatience opposite of patience, hate or indifference opposite of affectionate, etc.

Realizing the many positive traits you may have overlooked, as well as their opposites, may surprise you, but it likely won't feel completely foreign. They are, after all, connected to you. We all possess duality. Everyone has both good and bad traits simultaneously. Psychoanalysis refers to this as the "shadow self." Both the good and bad, the actual self and the shadow within, are you. Therefore, we must accept our duality. We should also examine if there is a sense of lack or an aversion underlying the things we like and pursue. These could be obstructive elements.

Since our current consciousness is created from the unconscious, they are essentially one and the same. Understanding this relationship allows us to easily map our subconscious. By reverseengineering our current consciousness to peek into the unconscious, we can discover what lack we're harboring. Only then can we truly hold the right intention.

According to Abhidhamma, often referred to as the psychological philosophy of Buddhism, the mind never exists in isolation. All "minds" are a combination of the mind and its contents (various cognitive information that constitutes the mind) and can never arise independently. Meaningless information stored in our subconscious starts to form groups when certain situations arise, forming meaningful patterns. Initially insignificant information morphs into meaningful data, emerging as "mind."

Ayurveda philosophy from India also describes the mind as made up of myriad points of thought, feeling, and sensation. These numerous points of the mind, without any specific shape, swiftly follow one another. When they come together, they form what we perceive as a "line" or "cognition."

When you randomly dot points on a page, they have no individual meaning, but viewed from a different perspective, they form a specific shape. The meaning of these dots always changes based on our perspective. Your subconscious will associate these points with shapes.

The emotions and thoughts that we perceive as "our mind" are a combination of information received from our sensory organs. Particularly, meanings are assigned to those that receive our attention.

We may like a woman who is beautiful to look at (visual information), has soft skin (tactile information), a kind voice (auditory information), a broad social network (personality information), and a kind heart (emotional information) because of the positive attributes weassociate with her.

However, the mind can also turn 180 degrees in an instant upon receiving new information. If it turns out that all her actions were a performance, her words and deeds don't match, she leads a double life, and she is a tremendous fraud, then any fondness for her will immediately disappear, replaced by feelings of displeasure and resentment. The mind changes like this all the time based on information. The essence of the mind is ultimately a cluster of information.

The *Dhammapada* speaks of the mind as something that "travels far and wanders alone. It is formless and resides in the caverns."

The mind's ability to travel far means it is not constrained by time and space. For instance, the moment we think of the United States, we can feel as if we are already there.

That the mind wanders alone means that at any given moment, only one "mind" exists—one by one! But it changes so quickly that it seems to be constantly shifting. For example, the moment we think of the United States, we might start to wonder about the cost of the trip, then go online to look for cheap flights, but then we might get distracted by a scandal involving a famous celebrity. And just like that, we may completely forget about the trip. So, while thoughts seem to occur sequentially, they are actually just a series of discontinuous bits of information.

The idea that the mind lives in the caves suggests that when specific information enters the dark recesses of our subconscious, it connects with related content, creating a meaningful cluster as one thought leads to another.

If we can never escape the erratic jumps of the mind, the creation of reality would be impossible from the start. However, if we realize that our thoughts are simply a collection of originally meaningless bits of information, we can then strengthen or dismantle those meanings as we wish.

When we catch ourselves fervently searching for cheap flights to the US, despite realistically not having the time to make the trip, and we ask ourselves,

"What am I doing right now?" that meaningless thought and action naturally come to a halt. We realize that it was just a result of having thought of the word "US" initially, and we can let it go without any lingering attachment.

Upon introspection, we can realize that a significant part of our thoughts, emotions, and actions automatically arise based on situational conditions. Conversely, if we fail to dismantle these snowballing thoughts in time, we end up wasting time and ultimately may lead ourselves to disappointment and frustration, blaming ourselves. Was it really a situation that warranted such self-criticism? It was, after all, just one insignificant thought from the start.

Let's practice dismantling our mind from negative emotions through the following separation technique. This practice can help quiet the stormy seas of our mind, bringing about a greater sense of calm.

Separation Technique

1. Recognize what thoughts are arising at this moment.
2. Examine why this thought has emerged and what reasons led to it.
3. Identify the initial topic or word that triggered this thought.
4. Acknowledge that the current thought (from step 1) is simply a derivative of one or two ideas (from step 3) and is inherently meaningless, originating from randomly surfacing ideas.
5. Mentally discard the answers (words) from step 3 one by one, observing how the current feeling changes.
6. If no significant resistance or negative response emerges after discarding the last word, it indicates that the separation of the mind is occurring effectively. If other reasons or emotions arise that hinder the separation, return to step 2 and reevaluate the beliefs held in your subconscious.

Defense Program

Let's revisit the brain. The human brain, which oversees all systems of the body, is functionally divided into three layers based on its developmental sequence: the neocortex, the paleocortex, and the brainstem.

The brainstem, known as the "reptilian brain" or "primitive brain," is the oldest in evolutionary history and directly manages vital life functions such as respiration and heart rate. The paleocortex developed with the emergence of mammals, often referred to as the "mammalian brain." It includes the limbic system and oversees emotions and stress responses.

Finally, the neocortex, also known as the cerebral cortex, consists of the frontal, parietal, occipital, and temporal lobes. It is responsible for higher-order mental activities such as memory, analysis, judgment, and language-based activities, hence sometimes referred to as the "human brain."

The paleocortex primarily manages our fear, a crucial survival tool that protects us in dangerous situations. It acts as a defense program throughout our lives. The neocortex also has its defense program, centered around doubt and judgment-based thinking.

Fear and doubt are special self-defense programs that humans have evolved for survival. They allow us to quickly discern what is safe and what is dangerous in various experiences. However, if these defense programs completely dominate our consciousness, they can hinder new experiences. Many of the defense programs inherited through evolution and ingrained in our collective unconscious are now unnecessary in modern civilized society. We hardly have to worry about immediate survival threats, like starvation or predator attacks. Yet, fears and doubts about survival can unexpectedly arise in our lives, prompting a defensive stance. While this can be helpful for survival, it can also be an obstacle to creating our desired reality. Therefore, we need to minimize unnecessary defense program activation, retaining only essential fear and doubt.

Mind is a collection of information, and the direction of these pieces of information dictates the direction of the mind. We possess what can be termed as "jury members of the unconscious" within us. Based on their experiences, they actively debate and decide on actions to keep us out of danger, then communicate these decisions to us. When we find ourselves internally debating between different actions, it's these unconscious jury members at work. However, they sometimes err, basing their arguments solely on their limited experiences. Even when presented with new facts and evidence, they might dismiss these as mere "excuses" and stick to their verdicts. They can be overprotective, often jumping to highly subjective judgments at the slightest suspicion.

These jury members in the mind utilize defense mechanisms, impressive rhetoric and thinking skills to argue in their favor. Some of the commonly used defense mechanisms are as follows:

Defense mechanism

Defense mechanisms are unconscious psychological patterns or behaviors that the ego uses to protect itself in threatening situations. These mechanisms distort or reinterpret reality to shield the ego from emotional harm. Here are some common defense mechanisms:

1. **Repression:** A primary defense mechanism against anxiety, repression involves unconsciously pushing unbearable thoughts, desires, and impulses into the subconscious.

2. **Restitution:** This involves self-imposed suffering to atone for perceived guilt, like believing one must dedicate their life to service to compensate for accidentally harming someone.

3. **Hostile Identification:** It refers to imitating someone you don't want to resemble, such as children who grew up under an abusive father and, despite hating him, exhibit similar violent traits as adults.

4. **Projection:** This involves attributing one's own flaws or undesirable traits to others. For example, being excessively harsh towards someone because they reflect your own weaknesses.

5. **Displacement:** This mechanism involves shifting emotions from the original source to a less threatening target. An example is a person feeling guilty about moral failings and compulsively washing their hands as a form of obsessive cleanliness.

6. **Denial:** One of the most primitive defense mechanisms, it involves refusing to accept reality or facts that are too difficult to handle, like a patient denying a cancer diagnosis.

7. **Rationalization:** It involves justifying irrational or inappropriate behaviors with logical explanations.

8. **Resistance:** This mechanism prevents repressed memories from surfacing,

sometimes even erasing the memory of the traumatic event itself, as seen in cases of short-term amnesia following a significant shock.

9. **Reaction Formation:** It involves acting in a way opposite to one's true desires to suppress them. For instance, a child who envies and resents their sibling might overcompensate by showing excessive affection and care.

10. **Sublimation:** This is redirecting unacceptable impulses into socially acceptable activities, such as channeling aggressive urges into sports or creative endeavors.

Our subconscious is filled with various defense mechanisms that can obscure the original information and events, making it challenging to recognize them for what they truly are. While these mechanisms aren't inherently bad, they aren't objective either. They can lead us to rationalize a distorted view of life, encouraging us to continue living as we have been, driven by our subconscious fears and limitations.

These defense mechanisms can make new experiences seem terrifying. When you attempt to do something that contradicts what your subconscious knows, it reacts defensively. This response can make it difficult to understand what we genuinely want, as subjective judgments can never represent the whole truth.

It's worth considering whether the things you currently desire might be influenced by subconscious deficiencies and wounds. The loud protests of your defense mechanisms could be steering your desires, suggesting that what you think you want might not be what you truly need or what would genuinely fulfill you. This realization calls for introspection and self-awareness to discern between authentic desires and those influenced by subconscious fears and defense mechanisms.

The Sweet Temptation of Scarcity

Jewels are precious and beautiful because not everyone can have them. If jewels were as common as stones lying around, no one would consider them precious or beautiful. If you have something no one else in this world has, you possess something with the highest scarcity. People will look at you with envy, and you'll enjoy their gaze with pride.

Who decided this value? The desire to be acknowledged, thinking, "I have something others don't, so I'm superior." Therefore, the value of scarcity doesn't lie in the scarce item itself, but arises when it aligns with your inner desires. If you're not drawn to it, no matter how many people call it a diamond, it's meaningless.

Just like the saying, "The grass is always greener on the other side," even if everyone has the same thing, others' possessions seem better. When we have what others don't, we feel immense catharsis. Conversely, if we lack what others have, we're consumed by defeat and inferiority. That's why we might always want something better than what others have.

But if you look at what we yearn for, most are values of scarcity created by mass media. Just like wearing the clothes and accessories of famous celebrities for vicarious satisfaction. Aren't you dreaming of living like the top ten richest people in the world, envying YouTubers who earn millions a month? Without considering if it's truly what you want.

We need to first figure out whether what we desire is truly what we want, or if it's a result of "scarcity" unknowingly influencing us. The standard of happiness is highly subjective, but we're brainwashed to believe pursuing specific things is the only way to a normal life. That's the problem.

You can be happy living differently from others. Actually, you might find happiness in discovering joys unique to you.

The Information Virus

Have you ever wondered what kind of thoughts you harbor and where you get your information from?

Dan Ariely, a behavioral economist, conducted an experiment showing how bias can affect taste. He offered students two types of beer at a bar: "Beer A" was regular Budweiser, and "Beer B" had a few drops of vinegar added. Without knowing Beer B's secret, most students preferred it. However, when informed of its contents, those who knew Beer B had vinegar were initially repulsed and chose Beer A. Surprisingly, those who liked Beer B first, even after learning about the vinegar, continued to prefer it, and some even asked for more vinegar. It shows how students who first experienced the taste without bias stuck to their choice, but those who had the preconception of "vinegar-tainted beer" couldn't appreciate its true taste.

If information can change our taste, can the body change through consciousness? Try this: Put your right and left index fingers together to compare their lengths. Then, separate your hands and focus on your left finger, repeating, "Grow longer, grow longer," and do the opposite with your right finger. After about two minutes, compare them again. See? The left one might appear longer. This shows we can influence physical changes just with thought. Consistently feeding certain information can lead to changes that seem impossible, whether it's health, appearance, or anything else.

The reason why early home environments can influence our future is that the quality of information obtained then directly impacts our lives. Mencius's mother moved three times to ensure her son was exposed to valuable information. We either accept the information we received in childhood or choose a life that defies it. A thrifty father might produce a frugal daughter, while another daughter, tired of such frugality, could be attracted to a man who is lavish and boastful. In reality, the trivial information of everyday life has significantly influenced us, collectively shaping who we are today.

"Why does information hinder the creation of reality?"

The answer lies in our subconscious. All information is stored in the subconscious as meaningless information, but certain situations trigger specific information linked to it. Then, these specific bits link up, surfacing from the latent state to dominate the conscious mind.

Information silently infects us. We act as if it's our own thoughts, but really, we're just responding to what's been put in our heads. If you think you're selectively and rationally accepting outside information, you're kidding yourself. We're constantly bombarded and brainwashed without even realizing it. I call this "virus information."

One day, I went shopping with a friend who saw a dress in a shop window and said, "That's pretty." It was a light blue checkered dress, totally not my style. She kept saying how much she wanted it, but I wasn't interested. A few days later, I saw the dress in a department store and impulsively bought it, pretending I had been eyeing it for a while. It wasn't my taste at all, but for some reason, it looked really nice that day. Even now, I don't understand why I bought it. I had been infected with my friend's information virus.

Like this, we're unknowingly governed by a myriad of information, believing it to be our own thoughts. A line from a song on a bus, a small flyer on the street, chitchat from a supermarket lady—all these bits of information are constantly piling up in our subconscious. Yet, all of it is just information—neither absolutely right nor wrong. If we don't realize this, we can't escape being slaves to information viruses.

Whether our subconscious leans towards positive or negative information depends on which type has more weight. This "information battle" is really a battle of energy. All judgments are made based on balance between positive and negative information. That's why we need to stay awake. If we live unconsciously, not aware of reality, we're nothing but slaves to bias and influence.

How can we escape the dominance of information?

Escaping the Information Virus

1. **Letting Go of Thoughts:** Since every thought is a combination of information, to free ourselves from the information virus, we first need to stop thinking. No matter how appealing a thought seems, firmly pause all those thoughts for a moment.

2. **Act Without Meaning:** Try doing something meaningless. Blurt out anything. Move your body randomly. Laugh like a fool. Don't try to find meaning in your actions.

3. **Follow Your Intuition:** If you suddenly feel the urge to do something, just do it. Don't think about what benefits it might bring you or what it means. This brings you closer to the truth of the moment. By not being swayed by thoughts and acting purely, you'll naturally learn how to move intuitively. Our subconscious is already perfectly programmed with what

we truly need, but random thoughts obscure it. Therefore, by stopping your thoughts and following your intuition, things you need will start happening smoothly and appropriately.

4. **Seeking Pure Information:** Pure information refers to the essence of the energy's information itself, existing beneath the perceptions overlaid by past experiences. Only by stripping away the overlaid information can you reach pure information through insight. However, pure information is always hidden, so you need to develop an inner vision to see it.

"Why Kill the Buddha If You Meet Him?"

Why do they say to kill the Buddha if you meet him? Did the Buddha do something wrong? No way! The mistake actually lies in the information we have. The Buddha is just a symbol, you know. It's just an image we've created from what we believe. Information is just that—information. Neither more nor less. So, we need to go beyond the Buddha image, to its essence.

Let go of all the information you've held onto and just see the person in front of you for who they are. Don't just replay the information you've always believed in. Instead, tune into the pure information that's revealing itself to you right now. Feel the unique energy behind what you're seeing.

If your old notions start creeping in, trying to interpret and judge, intentionally break that flow! Just chill and look at reality as it is. Notice any pure intentions that arise spontaneously, without being influenced by your reasoning, judgments, or external info. Then act on them!

"Act before you think!" That's the real way to find the answer without overanalyzing. We've always taken in information that suits us. But if you accept this moment as it is, the hidden reality starts to emerge.

What, then, is the real world?

Format and Reprogramming

Format

1. Is my thinking rational? → X
2. Is the decision I've reached after much thought correct? → X
3. Is information power? → X

Reprogramming

1. All information simply exists in its place, inherently meaningless.
2. Look beyond the information that meets the eye.
3. Escape the pitfalls of the information-filled world.
4. The only true belief one can have is in oneself, unaffected by the sway of information.

Errors in Perception

We often compare the human brain to a computer. This is because our brain processes information in a way that's similar to how computers do. We perceive an event, store the information derived from it as memories, and then these memories form the basis of our automatic thoughts. In psychology, this framework of automatic thinking is known as a "schema."

In 1934, the behaviorist psychologist John B. Watson conducted an experiment with his son, Albert, who was less than a year old. Whenever Albert touched a white rat, Watson would make a loud noise to startle him. As a result, Albert began to fear not just the white rat but any white object. Before this experiment, he had no negative thoughts about white objects. However, after repeatedly being startled in the presence of white, he developed a schema that "white is scary."

Once a schema like this is formed, it automatically triggers in relevant situations. Just seeing the color white can automatically cause a startle. But not everything white is bad, right? This automatic system reacts unconditionally, without considering good or evil or right or wrong. As convenient as it is, it also carries the risk of causing significant errors. This is called a cognitive error, meaning that an error in processing information is inevitable when we view reality through a limited schema. The types of cognitive errors are as follows.

Cognitive Errors

1. **Overgeneralization:** This involves applying broad general rules inferred from just a few experiences to unrelated situations. An example is someone who, after a first heartbreak, decides to live alone and never to love again.

2. **Selective Attention:** This occurs when you continue to conceptualize only the points you want to see, ignoring other information. This leads to a difficulty in understanding the whole situation and perpetuates distorted cognition.

3. **Personalization:** This is finding causes in oneself for external events without sufficient basis. An example is self-centered thinking where someone feels an excessive sense of responsibility to save the world or help others, often without concrete reasons.

4. **Catastrophizing:** This is consistently expecting the worst about the future. Common among pessimists and doomsayers, this error overlooks

that the world doesn't only flow in one direction. Eradicating global miseries like famine and war overnight is no less likely than complete disaster.

5. **Arbitrary Inference:** This involves making snap judgments based on schemas, without factual basis. An instance is when someone reacts with violence upon perceiving someone else's glance as hostile. Actions should be based more on realistic relationships rather than subjective emotions.

As we've seen, schemas often create cognitive errors and trigger incorrect automatic responses. We tend to use these schemas uncritically, leading to subjective interpretations of situations and often causing problems in life. Essentially, schemas are a product of the past. However, our lives are constantly changing. In this world, nothing is static—except perhaps the schemas in our minds.

The formation of a schema is crucial. Once it is established as valid, it triggers automatically in subsequent similar experiences. Such automatic reactions, often referred to as habits or stereotypes, are indicative of our brain's tendency to use less energy on stimuli categorized as "safe" according to our schemas. For example, we don't consciously focus on how we use chopsticks while eating or pay particular attention to each step while walking. Our everyday actions are processed automatically, leading us to overlook the subtle changes occurring around us. We only notice new information when significant changes or stimuli occur, enough to replace our past records. In other words, we are, in essence, living in the past, governed by our brain's automatic systems.

Let's consider someone who believes that meeting a beautiful woman is essential for a successful life. This belief may stem from observing examples where attractive women have formed happy families, raised successful children, and so elevated the family's status. It's like watching a drama unfold over the years. Observing the ultimate success of these characters, one might envy their lives and aspire to emulate them. This subconscious acceptance leads to the formation of a schema, a belief system based on that specific narrative. We all build our belief systems through such processes. However, these processes are prone to errors because they are based on limited experiences and can't be universally applied. If someone tries to impose this belief on others, including their children or peers, problems arise due to conflicting schemas.

"Meeting a beautiful woman is key to a successful life!"

"No, that's not true. What does appearance have to do with it? It's about capability, not looks."

So, how can we identify and reset these pervasive schemas in our lives? Discovering your schemas is easier than you might think. By paying attention to how someone routinely reacts and speaks about their daily life, you can discern their underlying schemas. Let's delve into your schemas using the following workbook.

workbook

Discovering Schemas

We can discover the schemas in our minds through questioning techniques. The following dialogue illustrates how schemas emerge in our thought processes.

Client: Recently, a customer called me and nitpicked for the most absurd reason. It was so frustrating and maddening.

Counselor: What made you feel frustrated?

Client: No matter how kindly I explained, the customer just stuck to their opinion and wouldn't understand what I was saying. I was on the phone for over thirty minutes with this incredibly stubborn customer.

Counselor: What did they say?

Client: They said they only wear 100 percent wool, but our product didn't feel like it to them. Despite the label clearly stating it's 100 percent wool, they wouldn't believe it. They even asked for the origin and manufacturing process. How am I supposed to know that? I just import and sell these products.

Counselor: Maybe their perception was correct? How can you be sure it's 100 percent wool?

Client: It's written on the label. That must be correct. How can I do business otherwise?

Counselor: Then why doesn't the customer trust the label? Why do they insist their feeling is right?

Client: Maybe they don't trust it because our products are imported from China? The price is a bit cheap, but would I sell fakes?

Counselor: Perhaps the customer had a similar experience before. Have you ever had such an experience?

Client: Well, now that I think about it, news always reports on fake luxury brands and food from China. Even department store items are often fake. I never eat food imported from China.

Counselor: Maybe the customer had the same thought. If you're so confident about wool, maybe their feeling about real and fake wool is right?

Client: That's possible. If they really like wool, they might know the difference between real and fake. I can even guess the origin of my favorite coffee beans.

Counselor: Do you now understand why they were so adamant?

Client: Yes, I understand a bit now. I take pride in the products I sell and thought I knew best about them. When the customer doubted the label, I was flabbergasted. But now, it makes sense. I should check with the factory. If it's not 100 percent wool as the customer says, I might need to rethink selling this product.

Counselor: It turned out to be an opportunity to reevaluate your product. That's something to be thankful for.

Client: Indeed, it is.

The client's mindset was shaped by a schema of pride and expertise in their product, leading to a failure to properly address the customer's concerns. Such schemas often cause errors in our lives. You can conduct this questioning process by yourself, using the above dialogue as a reference. Follow the steps to uncover your own schemas and understand the origins of your habitual thoughts and behaviors.

This is a practice for discovering the schemas I usually hold. Whenever possible, take time to follow the next process to discover what schemas you have. It will help you understand the origins of the tendencies in your thoughts and words that you unconsciously make.

1. Meet with things you haven't experienced before, like new photos, movies, places, or people. The fresher the stimulus, the better.

2. Record the emotions you feel in unfamiliar environments. Write down every reaction that occurs in your mind in detail.

   Example: "I'm not really into movies, but coming to the cinema, why are there so many people? Coming to the movies at this time. Well, I might be here, but it seems like these people really have nothing better to do," and so on.

3. Break down all the reactions you've recorded into individual sentences.

   Example: I do not like movies.
   I came to the cinema.

There are many people here. Coming here at this time (3 PM) made me think that these people have nothing better to do.

4. Separate each reaction into facts and subjective responses.

   Example: I do not like movies. (Response)
   I came to the cinema. (Fact)
   There are many people here. (Response). The number of people being "many" is also subjective.
   Coming here at this time (3 PM) made me think that these people have nothing better to do. (Response)

5. We should always be aware of the facts of each moment. Reactions are reflexes based on some schema in me. They are not real, just an illusion. Let's ask "why" about these reactions and discover what schema lies within.

   Example: I don't like movies. → Why? → I used to get scolded by my mom whenever I watched movies as a child.
   There are many people. → Why? → Isn't it a lot when about half of the place is filled?
   People coming at this time (3 PM) must be those with nothing to do. → Why? → 3 PM is naturally a time to work. Those not working at this time are jobless, aren't they?

6. All schemas create errors. Remove the identified schemas from your mind. They are just schemas formed by the environment you grew up in and have nothing to do with the current reality. Declare that you will let go of these schemas from this moment.

7. Rewrite your feelings in the initially unfamiliar environment. If the previous reactions come up again, intentionally look for different reactions, excluding them. Then, separate facts and responses again and look for schemas.

8. Only the facts are an accurate response to what is happening now. Remember this process and be aware of the subtle emotions you feel in every moment of life, checking whether you are creating distorted reactions based on schemas.

As we keep questioning, we eventually discover our deep-rooted schemas. These schemas may have arisen from wounds, fears, limitations, or worries accumulated over a lifetime. As long as these schemas remain, we're bound to commit errors in perceiving the external world. An objective observation of oneself reveals many cognitive misinterpretations unrelated to actual facts. Many of our schemas, unbeknownst to us, have been firmly guarding our subconscious. We live within a program where making new choices is impossible. What if your stubbornness, desires, goals, and dreams were all born from these misguided schemas? What if everything you firmly believed in was wrong? What if automatic thinking, influenced by distorted schemas, hinders the creation of reality and causes all problems in real life?

To create a desired reality, it's crucial to revise these internal schemas. The method to modify them is not too complicated. **Simply notice when a schema arises, pause for a moment, and then adjust your perspective to see the original information clearly.**

If we fail to notice and correct the errors of our automatic schemas as they occur, we might forever live without knowing what's going wrong. Yet, the reactions arising in our minds are so subtle, they're easily overlooked. I'll show you a way. By making mindfulness a habit, you can gradually modify your schemas. By observing each reaction that emerges in your mind, you strengthen your ability to control your thoughts. This task requires a lot of time, and the more schemas you can dismantle, the better. Always remember: the baseline for resetting schemas should be zero—staying free from all information is crucial.

Schemas form the core of our ego and prejudices. Therefore, removing narrow-minded schemas enables us to better receive intuitive and pure information.

Another method to break schemas is to act contrary to their commands. Deliberately do things you dislike and ponder, "Why did I hate or fear this?" Throw yourself into intolerable conditions. Remember what I said earlier? I deliberately sought out what people consider trivial ...

You might think confronting dislikes is tough, but it's not. Facing dislikes brings intense resistance, a reaction caused only by schemas. Curious about which schema is causing this? If you want to know, challenge the schema's command. Find the thing you dislike the most, close your eyes, and just do it. If you firmly believe in a habit, try thinking and acting oppositely (but don't break the law!). When you overcome significant aversion, you'll likely experience immense liberation and new insights. You'll realize you were trapped in trivialities!

Schema Modification Meditation

If you notice a schema emerging during your everyday life, try schema modification through visualization meditation. Follow these steps on your own.

1. Close your eyes for a moment and repeat to yourself: "The thoughts I currently hold are incorrect. They are my misinterpretations. They are mere illusions formed from my experiences, lacking real substance and meaning. Therefore, I let go of these meaningless thoughts and let them pass."

2. Visualize the schema as vividly as possible, then imagine it dissipating like smoke into the air. Realize that it is just a figment of your imagination and thus bound to vanish. For instance, if you have a schema that women must be beautiful, visualize an incredibly attractive woman. She is a hologram, captivating but inevitably going to disappear. Picture her slowly fading away until completely gone.

3. Realize that the notion no longer resides in your mind. It has dissolved into the air, leaving no trace behind.

4. Imagine a pair of transparent lenses that allow you to see things in their true form. Place these lenses over your eyes. With the lenses on, the true nature of things becomes clearer. Now you're not looking at a world distorted by schemas, but seeing the world as it truly was all along.

5. Open your eyes and immediately reconnect with reality. Observe your surroundings as they are, without allowing any distorted perspectives. What is the real appearance of things around you? How does the object of your attention look now? The previous schema you held has been modified. Moving forward, with a clear lens, you will see situations for what they truly are.

Breaking Schema Game

1. **Rename Familiar Objects:** Point at commonly seen objects around you and name them differently. For instance, point at a clock and call it "a dinner plate." Your schema might prompt you to use the usual name, but intentionally alter it. This separates the object from its conventional label.

2. **Invent Random Names:** Once you're comfortable with the first game, point at objects and give them completely new, nonsensical names. For example, look at a clock and call it "Zippatooey." Ignore all your usual language patterns. The function of an object isn't tied to its name, so feel free to rename it creatively.

3. **Speak Nonsense Words:** Normally, we combine familiar words to convey meaning. However, for this exercise, speak using meaningless words. Don't seek meaning in each word. Use sounds to express your feelings as if you're an alien. Even if no one understands, you're still conveying your intended pure information.

4. **Push Your Limits:** Do things you've never dared to do before. We all have self-imposed limits like "I'm too shy," "That's not my thing," or "I'm too scared to try that." Consider this a simple game. Without attaching meaning, do what you previously thought impossible for you. What happens when you cross a line you've never dared to cross before? Let's find out.

If you transcend schemas, that is, go beyond belief systems and automatic thoughts to face reality as it is, you'll discover an enormous amount of pure information. A world of infinite possibilities. At least, there'll be no false fears holding you back. The moment of creation should start from a world of infinite possibilities. Always welcome pure information with an open mind. Soon, a new world will unfold before you.

We've been living in illusions, not knowing what we truly want. Since it wasn't our sincere desire, we lacked the power to achieve it. Remember the O-ring test? We can't exert any power over what our subconscious doesn't truly want. Boldly breaking schemas prepares you to rediscover what you genuinely want. You must delve deep into your subconscious. If you can't overcome this programming, no matter how earnestly you wish and apply the law of attraction, you cannot create what you desire.

Format and Reprogramming

Format

1. I always accept new information and make the right decisions. → X
2. Cognitive errors only occur in specific situations. → X

Reprogramming

1. All information is fallacious.
2. Escape from information and embrace new pure information in every moment.
3. All thoughts and reasoning are fraught with errors.
4. The state of stopping thought is actually more correct.

Find the Initial Commands!

How can you be sure that what you desire is truly essential for you? Maybe you've been chasing vain ideals, feeling frustrated when they don't materialize. Have you been "creating reality" without even knowing what you genuinely want?

The failure to achieve what you desire could be entirely logical! The things you think you want may not be what you truly desire. Often, we live our lives mistaking the illusions fed by others as our aspirations. Hence, we miss out on understanding our fundamental desires—those that are genuinely different and deeply rooted within us. To attain what we currently want and need, we must look beyond our automatic thoughts and patterns, asking ourselves the fundamental question of what we truly desire.

Our actions reflect only the final commands processed by our brain, often significantly different from the original thoughts or commands. That's why it's crucial to discover the initial command. To find it, we need to backtrack from the final command that led to our actions.

You've always wanted that red sports car. But did you get it? If, after months or even years of intense longing, your wish hasn't manifested into reality, what could be the issue? Is it a lack of creative energy? Will it materialize if more time passes?

No, the real reason is that what you truly want isn't the red sports car. More precisely, the command for creation was incorrect.

You might be thinking, "But I really want that red sports car, what are you talking about?" However, while your conscious mind is fixated on the sports car, it might actually be that your subconscious desires something else entirely, like a beautiful house, a good job, or a happy family. If this is hard to understand, let me simplify it for you.

If you have a desire that just doesn't seem to be fulfilled, ask yourself why you specifically want that thing so much. The reason why you want a red sports car is important here. Why does it specifically have to be a sports car, and why red? Maybe your preference for red stems from a desire to be noticed, or it could reflect your passionate and active personality. There's a reason why you chose red over blue or yellow. If you discover a lack behind your choice, try to fulfill it. Then, you might choose a different color, or your obsession with the color might disappear altogether. Why does it have to be a sports car? Is it because you want to enjoy the feeling of freedom and exhilaration while speeding? Or are you feeling trapped and need an outlet?

Or, is it because you want to show off a sleek sports car? What's the reason behind this desire to show off? Is there a specific person you want to impress?

The red sports car is merely a symbol. Behind this symbol, there are various insecurities. If these insecurities are somewhat fulfilled and you are no longer influenced by that energy, you might find yourself desiring something other than a sports car. What you fundamentally want is to find that initial command! Discovering this initial command is the key to creating reality.

We can have anything, from red sports cars to grand palaces, immense honor, or blissful families. Anything is possible in this infinite universe, as long as we follow the laws of creation and place our orders correctly.

"You know, now that I think about it, I was actually just lonely. I wanted love and attention, and I comforted myself thinking that having a red sports car would make things better."

This approach requires insights to identify what we genuinely want and what we lack in our subconscious. When the hidden initial command is revealed, only then can we activate the energy of creation and resonate with the universe. This is the fastest and most accurate way to manifest our desires.

As a result, the reality that best matches the desires of my subconscious is created. Even though my conscious mind wanted a red sports car, if I realize that what my subconscious truly desired was love and attention from people, I might end up meeting the love of my life instead of getting the sports car. At that point, I probably won't even be interested in the red sports car anymore, right? Only when the fundamental desire is satisfied can we experience happiness and satisfaction beyond what we had imagined.

This is what true creation is about. Creation is happening all the time, everywhere. Reality is precisely created as led by our consciousness. It's just that we don't recognize it as creation. The reason why it's hard to recognize creation is because the reality and the desires on the surface of our consciousness are different. However, within the process of creation, even phenomena that seem completely different from our thoughts and reality are actually steps towards what we truly want. If you can discover your initial command, the thought and the created reality will align, leading to a direct and uncluttered act of creation.

Write down your wishes and deeply introspect until you can answer the question, "Is this what I truly want?" The thing you truly want might be completely different from the symbolic object that comes to your mind. The first step to fulfilling your desire is to clear away the superficial froth and find the initial command.

workbook

Finding the Initial Command

1. What is it that you ardently desire?

2. What are the reasons behind this desire? Write at least six of them down as they come to mind and add any additional reasons you discover.

3. Revisit these six reasons and ask yourself, "Is this truly what I want?" Remove any that are not genuinely for your happiness but are rather notions imposed by others or society.

4. For the reasons that remain, ask again, "Why do I want this?" When you can't find any more logical reasons, the feeling that remains in your heart is your initial command. Compare it with the final command you have been following. If they differ, re-adjust your goals to reflect your initial command as much as possible.

5. What is your Initial Command?

Before starting to create your reality, go through the process of identifying the initial command for everything you currently desire. Check if you truly want these things. What you have been longing for may not have been what you truly desired.

We often wish for changes and feel frustrated when they don't materialize. From today, let's act with a clear understanding of what we want. If you're still unsure about what you want, keep repeating these exercises until you find it. Analyzing "yourself" is essential to activate the Zero System for reality creation.

“

Format and Reprogramming

Format

1. Are my desires and wishes correct? → X
2. Do I know what I truly want? → X

Reprogramming

1. I may not fully understand what I truly desire.
2. My desires could be endless.
3. Sometimes, letting go of desires is necessary for new progress.

”

Inner Mind Map

Even though tangled schemas and automatic systems contribute to our cognitive errors, what's more serious is that we still don't fully understand how our subconscious system is formed. Research on human consciousness is still in its infancy. How can we claim to be the true masters of our lives if we're helplessly swayed by our thoughts and feelings? I refer to the subconscious as the "Inner Mind." It's the underlying "inside" information that forms the foundation of our current consciousness. Having helped many people with healing, I've noticed that most people share similar patterns in their Inner Mind. Although the way these manifest can vary slightly due to postnatal factors, the intrinsic Inner Mind of humans is fundamentally similar. I've completed the Inner Mind Map through various experiences and inspirations. This map represents the flow of our subconscious, and understanding it can expedite the process of finding our initial commands. It also helps in understanding what others desire and why they suffer.

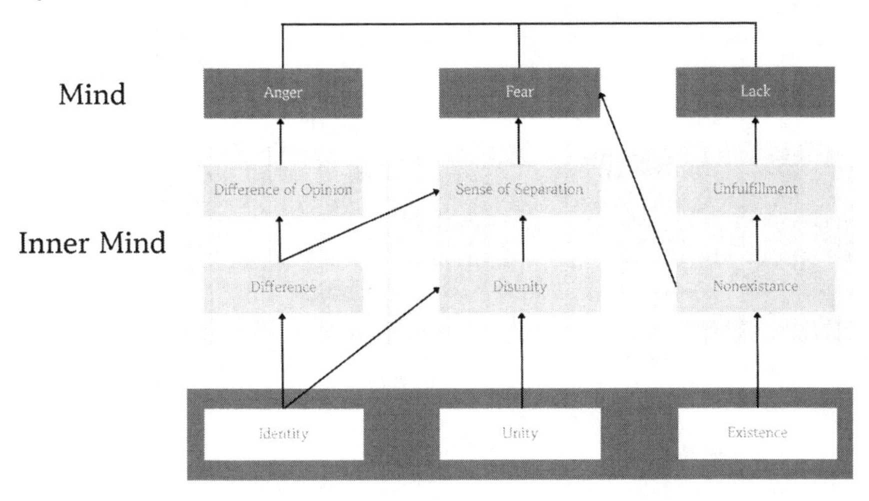

The key emotions that often trouble us in daily life boil down to anger, fear, and a sense of lack. These three emotions are the root of all negative feelings, and ironically, they are interconnected and dependent on each other. Underneath these emotions in the Inner Mind, there's a core that triggers them.

Anger arises from the difference in opinions between oneself and others. It's a friction caused by the discrepancy of not being aligned. When do we usually get angry? For instance, you might feel anger when your boyfriend doesn't do what you want. "Why can't he understand my feelings?" Or, "Why is my son always playing computer games instead of studying?," and, "The nagging of a mother to study can be exasperating." What is the real issue here? Is it because others don't understand us, or is it because things don't go our way? Why doesn't the world conform to our desires? People often view the world from a self-centered perspective, and many problems in life stem from this mindset. But is it wrong to think from our own perspective? No, it's a completely natural behavior. In our deep subconscious lies a core called "identity." **We subconsciously find comfort in things that are similar to us and seek to stay within that comfort zone.**

Fear arises from the worry that others might leave us or from the fear of loss. The feeling of being separated from others always instills fear in us. We unconsciously desire to become one with others. I call this the "oneness" attribute of the Inner Mind. Due to this attribute, we feel secure when we meet people similar to us. Subconsciously, we strive to belong to a group that resembles us. The larger the group, the more comforted we feel, knowing there are others like us. On the other hand, being alienated from the group can engulf us in immense anxiety. Humans are social creatures who find it difficult to live alone. We also fear the disappearance of our own life or that of others. People are scared when watching horror movies, right? Why? Because it seems like someone might threaten or even kill us, like in the movies. Why is death scary? **When we die, we can no longer exist in this world. Or, if someone else disappears, they can no longer be with us. We have a great fear of "non-existence" because we always want to "exist."** Therefore, when our state of existence is threatened, it results in immense fear and aggression.

The Inner Mind, which creates anger and fear, also gives rise to desires for change from any current unsatisfactory situation. **We don't usually have additional wishes for things we already possess and are content with. Why desire more from something we already have? All of our desires stem from the lack of existence.**

Numerous inner minds ultimately generate our myriad desires. Our life, endlessly desiring, originates from our subconscious. If most of our desires stem from a sense of lack, let's delve deeper into the initial stage where this deficiency began. What could be at the very core of our Inner Mind? Interestingly, at the very bottom of the Inner Mind core, there exists the notion of **"Being One in Existence,"** symbolizing unity, oneness, and existence.

Oneness

This is essentially being connected to everything, the very consciousness of the universe itself. Isn't this what we truly desire?

Disappearing "I" at the Center of the World

Have you heard the saying that the center of the world is always "me?" It's true, isn't it? In the micro world, everything is defined by my observation, and without the observer, that world might as well not exist. If you don't exist, what does it matter how the world turns? This world only has meaning when you exist!

Every scenario in the life you experience is undeniably about you. We constantly choose our lives at every moment. "I" am very important. But to initiate creation, "I" must disappear. This isn't about summarizing all the life I've lived and doing nothing. I am always the master of my life. And I need to be here. This is my world. That's unchangeable. So, what does it mean to make "me" disappear in my life? Here, "I" refers to all the images assigned to me throughout life. It's about the concepts that are recognized as "me." It's all the stories that are expressed as "I am a certain type of person," and the adjectives that define me. In reality, we're surrounded by various physical elements that constitute our environment. All human relationships, fame, wealth, houses, clothes form a worldview that identifies "me." The elements that define and modify "me" are like "adjectives" that grant value and meaning to my existence.

I am a doctor. (Profession): I am someone who knows how to treat people.
I own a big house. (Wealth, House): I am wealthy enough to own a big house.
I am a bestselling author. (Fame): I am famous and receive a lot of respect.
I have many friends. (Human relationships): I have a large network to utilize.

We start to construct "me" through physical elements that define us. That becomes our age and the world we exist in.

I am a doctor, own a big house, recently published a book that became a bestseller, and hence became very famous, and thanks to that, I've got to know many people!

That's me. But is that really me? The true me isn't the things that adorn me. The truth is, I am simply I.

I am a doctor.

The statement "I am a doctor" is an equivalence (=) expression, meaning that I am equal to the term "doctor." It suggests that I can be expressed as synonymous with the word "doctor." However, in reality, they are neither identical nor equivalent. The only word that can express the truth of my existence is "I." Being a doctor is not the entirety of who you are.

What you need to understand is that this "I" is the pure "me," devoid of any modifiers. Often, we live under the illusion that the many adjectives attached to us define who we are. But I reiterate, that's not the truth. I am purely "I."

If I want to be a doctor, I can become a doctor.
If I wish to be famous, I can become famous.
If my aspiration is to be a painter, I can become one.

It's only after a complete self-awareness that all the adjectives that are attached to us can become part of who we are. The images that adorn "I" are infinite. I can be anything, right? Therefore, I can freely attach any sentence behind it. The moment I attach a sentence, I become that.

I am a painter!

I am a being capable of endless transformation. As soon as I attach the adjective I desire, I become that. That's why I am just a pure "I," unable to be defined by anything.

Who am I?

That's exactly what "I" is! So, who is "I"?

I am I

I am simply me.

Earlier, I mentioned that the current "me" needs to disappear, right? It means to strip away all the adjectives attached to "I." When you discover yourself without all those modifiers, that's when your life begins to be recreated from within. The moment "I" disappear from the center of my world, I meet my true self. When I return to my place, that's when I can begin to recreate the life I've desired. You lack nothing, and you are not deficient in any way. All the lack within you is not you. And the "you" who desires something is ultimately not you either.

You are simply you.

Creation must start from this state of complete "I." Then, you can command the universe as you are. You can truly obtain what you desire!

Oh my god! What was this "me" that I thought I knew all this time?

workbook

Recognizing I am I

This process is about removing all the illusions we've known as "me" and recognizing the true "me." I am "me" who can't be defined by anything. Let's practice removing all the adjectives that make up "me."

1. Write down all the words you or others have used to describe yourself throughout your life. Examples: "don't have money," "tall," "overweight," "doctor," etc. I am ______________.

2. Review the words you've listed and think about whether they truly represent "me" or if they are just adjectives describing "me." If they are merely descriptors, consider whether "me" can exist without them. Then, boldly erase them one by one. For example, if you've thought of yourself as lacking money, is that really defining "me?" Or is it just an adjective? Ask yourself these questions. Having or not having money doesn't define "me." It was just a misconception. If that's the case, erase "don't have money."

3. The process of erasing each word is a journey to discover the errors in living under the misconception that adjectives were "me." Once this process is completed, you'll be able to feel the existence of "I am I," where nothing else exists. Now, feel the complete "me" with no adjectives, and replace the blank with "me." I am me.

4. Remember this feeling. I am I. The complete "me" with no adjectives.

Chapter 3. System

The Zero System of The Universe

Tao told is not eternal Tao.
The name named is not the eternal name.
The nameless is origin of heaven and earth.
The named is mother of ten thousand things.
Desireless, you see essence.
Ever desiring, you see only form.
Mystery and manifestation emerge from the same source:
Darkness.
Darkness within darkness.
The gate to all mystery.

— *Tao Te Ching*, Chapter 1

What Is the Zero System?

The Zero System refers to the mechanism and structure of the universe that encompasses the mystery of creation and dissolution. While "zero" might commonly bring to mind the number "0," it signifies more than just emptiness or nothingness.

One day, I realized that although all lives take different forms, they share a commonality: they are governed by the law of zero. This insight wasn't unique to me; it's a truth that many wise figures throughout history have taught.

Zero is simultaneously the smallest and the largest concept. I use the term "system" with zero because everything in the universe continuously influences each other, existing organically within this vast law of zero. I discovered that our lives operate under the same principle. The physical laws of the universe apply to life itself. They are the essence of the creation I experienced. I compiled these laws into what I call the Zero System.

Reality creation naturally occurs within the Zero System, but isn't an end in itself. The purpose of the Zero System is to enable us to live as true creators. As I shared the principles of the Zero System with people they too began to experience creation in their lives, confirming that the Zero System's principles apply to everyone.

The ability to create reality is a natural and inherent power. The deeper our understanding of life's true purpose, the more powerful our ability to create becomes.

I believe you too can harness this power.

Zero Law 0 – Zero and Infinity

Before delving into the Zero Law, let's reconsider the special number "0," commonly perceived as "nothingness."

Our numerical system today originated from Roman numerals. The Roman numeral system had seven symbols—I(1), V(5), X(10), L(50), C(100), D(500), M(1000)—and all numbers were represented by combinations of these symbols. In ancient times, large numbers weren't necessary for daily life, so Roman numerals sufficed. However, as the need to calculate larger numbers increased, an Indian mathematician invented the concept of zero (0). With the invention of zero, a system using ten Arabic numerals from 0 to 9 was established to express all numerical concepts, significantly advancing mathematics and science.

Initially, zero was used as a symbol to fill the empty space between numbers. For example, to differentiate between 1 2 and 12, 1 2 was written as 102. In Sanskrit, zero was called "*sunya*," meaning "emptiness." But emptiness here signifies more than just a void; it's a special space where any number can be placed. Ancient Indians, realizing the truth that all things lack a distinct essence, named the infinite "emptiness," which seems like nothing but exists as a reality behind everything, sunya.

When the ends of the number 0 are joined and twisted, it forms the symbol for infinity (∞). This symbol, representing the concept of no beginning or end, symbolizes eternity and the endless possibilities of the universe. Zero and infinity, though appearing different, are essentially the same. While interpretations can vary based on perspective, their essence remains unchanged. Zero appears empty but symbolizes the space of the universe, filled to the brim. From one direction, it seems nonexistent, but from another, it's infinitely filled. Therefore, zero signifies both nothingness and infinity.

Now, let's explore the astonishing principles of zero (0) that can be observed around us. The Zero System consists of nine Zero Laws, ranging from 0 to the sideways 8, representing infinity. These nine principles of the Zero System will reveal to you what the consciousness of a creator truly entails.

Zero Law 1 – The Law of Relativity

The world is made up of relativity.

Creation and Annihilation of Particle-Antiparticle

One of the great discoveries in quantum physics is that every particle exists in tandem with an antiparticle, possessing opposite properties. When a particle and its antiparticle meet, they annihilate each other, returning to nothingness. This represents the principle of existence and extinction inherent in all beings in nature.

For a property to manifest in the universe, its opposite property must be presupposed. Darkness must exist for us to comprehend light, and the concept of length arises only in the presence of shortness. In other words, no property can manifest in the universe without its relative counterpart.

For example, one cannot accurately gauge the temperature of water solely by touch. However, upon encountering colder water, we realize the previous water was warmer—understanding attributes relatively. When two different temperatures of water mix, their individual properties vanish, returning to a state of nothingness—an annihilation of polar opposites.

The Zero System is also based on this principle. Zero is not just emptiness but is closer to the fullness of everything. Disappearance or absence is just a phenomenon perceived through the narrow lens of human perception.

When particles and antiparticles meet and annihilate, they emit high-energy light. When matter vanishes, its dual properties disappear, and something new is created in its place. Light (photons) can also collapse and create new particle-antiparticle pairs. Photons, having no charge, comply with the law of conservation of energy, therefore necessarily creating electron-positron pairs with equal and opposite charges.

Particle + Antiparticle → Annihilation → Light (Energy) → Collapse → Particle-Antiparticle pair

This is the fundamental principle of creation. It may seem like matter disappears out of nowhere, but nothing truly vanishes in the universe. It merely transforms from opposing poles into other forms of energy and back into matter. Creation always arises as a pair of opposites.

In physics, the void where the basic materials return after the collision of these relative properties is called the Zero Point Field. In this realm of nothingness, particles and antiparticles continuously emerge and annihilate in a state of quantum fluctuation.

Thus, matter and energy constantly shift forms, emerging and vanishing. Everything that exists embodies its opposite relativity. The secret of creation lies within this relativity or duality. By keeping in mind that everything in our world exists on this grand mechanism of duality, we can find hints for creation within it.

The Harmony of Yin and Yang

The Eastern view of the universe is formed from the state of *wuji*, which refers to a state of emptiness that contains all relative properties. Wuji divides into two, forming the *taiji* of Yin and Yang, and within taiji, the latent Yin and Yang emerge and differentiate, creating the myriad things in the universe.

In the *Dongui Bogam*, a traditional Korean medicine book, there is a mention of a special remedy called Yin-Yang Tang, used when one's vital energy is low. This remedy is made solely from water. You fill two-thirds of a cup with hot water and top up the rest with cold water, utilizing the principle where hot water rises and cold water descends. This creates a new flow within the cup, combining Yin and Yang to generate a special energy that replenishes vitality.

When Yin and Yang, the opposing properties, meet, an automatic movement to harmonize occurs. All relativity contains this special force. Humans contain masculine and feminine forces, and are instinctively attracted to each other. During intercourse, a special energy (orgasm) is experienced, which is unique and can even create new life.

The meeting and harmonization of opposite energies are a special mystery of the universe and a gateway to new worlds. Such harmony occurs when there is a perfect balance, an exact cancellation of opposite properties, returning to Zero. If one side's energy is excessive or lacking, it's not harmony but an invasion or overflow. True equilibrium, where both sides precisely counterbalance each other, is the moment when new qualities can emerge in the world.

As discussed in Chapter 2, before longing for something, we first need to understand its absence, its opposite. Also, to solve a problem, we should try the exact opposite action. This underscores the importance of relativity.

The essence of Zero Law 1 is that when opposite energies precisely cancel each other out, they vanish, and a miraculous harmony occurs. This is the starting point of creation.

Earth's Zero Magnetic Field Zones

The phenomenon of special occurrences when two opposing properties meet can also be found on Earth. Our planet can be considered a giant magnet, with its magnetic North and South Poles. The Earth forms a vast magnetic field around its axis, protecting us from the powerful solar winds that emit electrons. Humans, inevitably influenced by this vast magnetic field, also have magnetic-like properties. Our bodies are often represented with the crown of the head as the N-pole and the Yongcheon point (center of the soles) as the S-pole. The flow of biomagnetism is said to promote health, whereas an imbalance can lead to problems. Friedrich Anton Mesmer, known as the father of hypnotism, proposed the theory of animal magnetism. He believed in an invisible physical force influencing the human body and used magnets to treat patients.

Within Earth's vast magnetic field, there exist areas with no magnetism, zero magnetic field zones. These areas emerge where the magnetic properties of the N and S poles neutralize each other, creating a space with "no polarity." Zero magnetic field zones are primarily found near major fault lines on Earth. Measurements of the Earth's crust in these areas have revealed near-zero fluctuations in the geomagnetic field. Faults are geological structures that appear where two opposing forces meet and balance each other out. It seems that the principle of returning to a state of nothingness, or zero, when two opposing forces meet, is also evident here.

These "zero magnetic field zones" are often known as "healing spots" or "auspicious places." When opposing energies balance each other with equal force, they transform into a new zero energy, and with the disappearance of relativity, new creation occurs. The formation of a zero magnetic field is also known to increase negative ions, which reduce fatigue and stress, and boost immunity. This explains why our ancestors emphasized the importance of auspicious places. Many such sites, often surrounded by mountains (yang) and water (yin), exhibit a harmonious circulation of yin and yang energies.

The harmony of relativity is a special magical gateway that unveils the mysteries of the universe. By viewing all life through the principle of relativity, we can find a simple way to resolve issues.

Every law, whether in the micro or macro world, applies the same principle.

Zero Law 2 – The Law of the Middle

Always maintain the middle in relativity.

Zero Point

The Zero Point is literally the zero (0) point. On a number line, the Zero Point is a space where no number intervenes, representing the boundary between positive and negative numbers. It is precisely at the center of the vast world of relativity.

Magical occurrences happen at the Zero Point, where the opposite poles of relative properties meet in balance. Right at the point where different relativities meet lies the secret door that opens to the mysteries of the universe. However, to find this door, do not lean toward any side.

A Zero Point without bias is not just a simple median value or a mid-position. For instance, in a seesaw, the balance point always changes depending on the weight on each side. To balance, the heavier side needs to sit closer to the pivot. The location of the middle point varies depending on the situation. Mass is relative but also absolute. Though it may seem different at a glance, it's the same in terms of the absolute mass value of the universe.

The Zero Point in reality reflects all changes and is constantly in motion. The Zero Point of the universe is different from the zero point of statistical common sense. It exists not from a human perspective, but to balance the entire universe.

Therefore, to find the Zero Point, one must understand the perspective of the universe beyond the human viewpoint. Just as one needs to shift from their own perspective to that of a creator. Then, we can understand the mysterious mind of the creator and gain the corresponding power.

The Space Between Spaces

The Zero Point exists in the "interstitial space" between one property and another, representing not just a single point but a certain space, like a dimensional gate. It's a "gap" that connects the physical world and the infinite universe.

As mentioned before, zero (0) is a symbol with the same meaning as infinity (∞), and so the Zero Point connects us to a potential space containing parallel universes of infinite possibilities.

Accessing this interstitial space is essentially connecting to the nature of the universe beyond time and space. At that moment, a world of infinite energy, transcending the physical laws of Earth, unfolds. I often describe being in this state as "entering the realm of zero."

In the Japanese animation *Cyber Formula*, which is about car racing, there's a concept known as "Zero Zone" among the racers, considered the ultimate realm. In the Zero Zone of the animation, when the driver accelerates to the limit, they momentarily transcend their senses and can see a few seconds into the future, experiencing a phenomenon where time seems to flow very slowly. Racers who enter this Zero Zone exhibit driving skills beyond imagination and dramatically win the race. In reality, athletes after long practice reach a state where they compete beyond space and time. They see a fast-approaching ball as if in slow motion, discerning its minute rotation and timing, or foreseeing an opponent's punch to evade in advance. In this state, the concept of time as we usually perceive it flows completely differently.

Tibetan Buddhism speaks of Shambhala, a legendary city inhabited by enlightened sages somewhere in the Himalayas. It's said that when the world nears its end, engulfed in war and desire, the army of light from Shambhala will emerge, defeat all evil, and ultimately triumph, leading to eternal liberation for people. Many have set off in search of this mythical city, but no one has ever found it; it's only known through legend. When asked if Shambhala truly exists, Tibetan monks reply, "Shambhala definitely exists, but not everyone can go there. Only those who are enlightened can see it. Everything exists within your mind." Shambhala is not a physical city. It exists in another dimension, always present here and now, just beyond our 3D perception. With a different sense and consciousness capable of perceiving other dimensions, one might encounter it.

In the movie *Doctor Strange,* there's a scene where Doctor Strange, after strenuous training, opens "portals" to travel through time and space. Remember how he spins his hands to enter another dimension? Such a portal is like an entrance connecting to Shambhala. This special point always exists in this space. If you are prepared, you can meet it.

We must draw the energy of creation from the Zero Point. For now, just remember that there is a special point called the Zero Point in our space. I'll talk about how to connect to the Zero Point later.

We should always strive to stay at the Zero Point, the point of balance. So where can we find this Zero Point?

Zero Law 3 – The Law of Contradiction

Zero Point is found within contradictions.

Existing Between Judgments of Good and Bad

We like to make clear judgments about right and wrong, taught to always pursue what is good and avoid what is evil. However, as one energy grows stronger, its opposite energy also intensifies, following the laws of the universe.

Everything possesses its corresponding opposite. Although we reject wrong thoughts and evil actions, they are also attributes created by our pursuit of good.

The boundary between good and evil is incredibly vague. We subjectively distinguish them based on the situation. As the Korean proverb goes, "The arm bends inward." We tend to view what is familiar to us more positively. Biased judgments inevitably move our coordinates closer to ourselves, distancing us from the Zero Point.

If you have thoughts like "this is good" or "this is right," your mind is already leaning to one side.
If you think "that is evil" or "that is wrong," your mind is again biased.

Past Experience → Thoughts and biases that inform our judgements → Judgement → "This is right" or "This is wrong"

We already possess thoughts that skew our judgment!

Judgment itself is a result of choosing one side. We unconsciously lean towards one side based on past experiences. Let's pause our thoughts for a moment.

When we stop thinking, we can reconsider the duality that emerges at the Zero Point. By realizing our one-sidedness, we can re-examine both sides. Then we'll understand why good must be good and why evil must be evil. We'll discover their differences and commonalities.

In ancient China, a merchant sold spears and shields. He boasted, "This spear can pierce any shield, and this shield can block any spear." A passerby asked, "What if your spear is used against your shield?" The merchant was at a loss for words.

A skeptical bystander considered the merchant a liar. However, an old man watching quietly handed the merchant a stack of money, saying, "How can there be a shield so strong and a spear so sharp? I know weapons, and both are truly remarkable. I'll buy both!"

Instead of wasting time debating which is stronger, isn't it wiser to appreciate the excellent qualities of both weapons? We usually think choosing means picking one of two options. But why must it be so? Can there be another choice? If we ponder a little deeper, we'll realize that acknowledging and harmonizing the differences of both is much better. Only then can new energy be created.

When we mediate between arguing friends, not taking sides but pointing out, "You're right about this, and you're right about that," they see qualities in each other they hadn't noticed.

"Both seem right to me, with such brilliant ideas. If combined, they might make the best approach. What do you think?"

Only an observer outside the fray can see both sides of a situation. Observers accept both sides and return the final decision to the parties involved. That's when they encounter another world they hadn't realized existed, leading to a new choice and creation.

We often repeat choices confined by our preconceptions, leading to monotonous lives. The reason we struggle to identify our own problems is because our thoughts and judgments are biased towards our perspective.

If you must choose or decide, pause all discriminations and judgments. Stay at the Zero Point, the boundary and center of relativity. Soon, a new perspective and possibilities will unfold.

Throughout life, we face countless crossroads, requiring the best judgments. From now on, when you need to judge, pause and stand at a value-neutral point, viewing both sides equally. You'll find a new direction, unlike previous choices.

There's no absolute good or evil in this world. Everything is right from different perspectives. The true answer lies in the unbiased midpoint.

"Everything is right" is not indecisiveness, but wisdom. The world is full of contradictions, and often, these seemingly nonsensical contradictions are closer to the truth.

Nonsense Makes Sense

Sometimes, in life, events that defy understanding or common sense occur. What could possibly be the right answer? We immerse ourselves in finding solutions, only to often discover that something entirely unexpected is the answer. Sometimes, there may not be an answer at all.

"How can you put an elephant in a refrigerator?"

Most people, perplexed by this question, respond with queries like "How big is the refrigerator?," "What's the size of the elephant?," "Can the elephant do yoga?" Despite their efforts, a feasible method seems elusive.
So, how can you put an elephant in a refrigerator? The answer is surprisingly simple!

1. Open the refrigerator door.
2. Put the elephant in.
3. Close the door.

When I reveal the answer, people often scoff, "That's impossible. What are you talking about!" Of course, it's impossible by common sense. But is there any other answer?

In *The Little Prince,* there's an illustration of a boa snake that swallowed an elephant. Adults saw it as a hat, but the little prince kindly drew the elephant inside the snake, emphasizing, "It's a boa snake that has swallowed an elephant." Still, the adults dismissed it as nonsense, insisting on their perspectives.

Was the little prince's thought "wrong"? In the world of adult common sense, a boa snake swallowing an elephant is impossible, just like putting an elephant in a refrigerator. But in the little prince's drawing, there's no problem at all, because everything is possible in the universe.

Let go of your need to find the right answer. Abandon all the criteria that determine feasibility, and just observe the phenomenon as it is. Often in life, what seems nonsensical may be the answer.

Sense can solve matters of sense. On the other hand, nonsense can only be solved with nonsense. In the realm of nonsense, nonsensical answers are the appropriate ones; in other words, they make sense.

Applying the standards of logic and common sense to things that cannot be proven is futile. Reality creation is similar in context. It's not about logic or reason, but about miracles and mysteries themselves.

Can you explain how the universe came to be, unfolds, and disappears? To prove this, you must enter a world beyond logic. The universe exists in more dimensions than we can imagine. Understanding the universe requires a lot of imagination.

Efforts to explain everything logically and questions like, "Can it really be achieved?" are actually useless. Before starting reality creation, we must first believe that even seemingly impossible futures can indeed be realized. The second is a perspective that views situations with an open mind!

Like the simple belief and perspective that an elephant can be easily put into a refrigerator. When such nonsense is accepted as sense, we experience astonishing things. The universe operates in ways that are entirely different from our common sense.

The logic and common sense of science are limited to what humans can observe. In other words, what cannot be observed falls outside the realm of logic and common sense. Also, common sense is subjective, varying with time and culture. Therefore, it's essential to recognize that there's no such thing as "absolute common sense."

When we stop viewing the world through our predefined frameworks, we finally encounter the real world. This is the Law of Contradiction. If we start looking at the world from a completely different perspective, we'll realize that anything is possible in this universe.

Think about it. If we willingly embrace the infinite possibilities beyond common sense, how much broader could our life experiences be? If we confine ourselves to our limits every day, we'll miss out on the diverse and incredible events in life. Accepting the incompleteness of our common sense and information and adopting a broad capacity to eagerly embrace new information is necessary. Only then can we encounter a higher level of consciousness and infinite energy.

Zero Law 4 – The Law of Non-Emotion

Always maintain equanimity from all emotions.

The Doctrine of the Mean

Maintaining the Zero Point equates to "staying in the middle." Not swaying to one side and keeping the midpoint is the essence of the zero system.

So, how do we maintain this position in life?

The Statue of Justice holds a scale in one hand and a long sword in the other. She balances the weight of the crime on one side of the scale and its corresponding retribution on the other. If these balance out, she deems it "fair"; if not, she uses her sword to behead the culprit.

Her judgment allows no sympathy or compassion. Since she is blindfolded, she cannot see who committed what crime. She judges solely based on whether the scales are balanced.

There's a saying: "Heaven is emotionless." Heaven, indifferent to human wishes, sometimes brings drought, sometimes floods. Humans may resent this apparent apathy, but that's merely a human perspective. Nature merely enacts what needs to happen, dispersing what needs to disappear. Emotions don't exist in natural laws.

Confucianism teaches the highest virtue as Doctrine of the Mean, emphasizing maintaining a middle ground in life. The Doctrine of the Mean advocates a state of not being swayed by emotions like joy, anger, sorrow, or pleasure. Such absence of emotional fluctuation is the key to staying in the middle, the Zero Point.

Emotions are feelings or states of mind that arise when facing certain situations. As discussed before, the mind is a collection of subjective information from our experiences. We know nothing of what we haven't experienced, merely speculating based on subjective judgments. However, this approach won't find the balanced midpoint. Insisting that your current position is the center of the universe, while actually being biased, doesn't align with the universe's power.

To let go of subjective views and judgments, we must choose not to be swayed by emotions. Abandoning notions of right and wrong, good and evil, likes and dislikes, we should find our place only at the Zero Point, free from any bias.

Maintain emotional balance even in situations that seem favorable or challenging. As the saying goes, "Don't count your chickens before they hatch." Premature excitement or worry about unfolding events can lead them to drift away. By balancing emotions, you can genuinely embrace what's coming. I first grasped this principle in middle school; remember the story about the exam?

Emotional balance is a key that influences the creation of reality!

Sometimes, the hardships you face may seem too harsh. You might want to blame the heavens, wondering why life is so tough for you while others seem to have it easy. But even if reality appears that way, you must maintain composure without being swayed by emotions.

As you continually seek the Zero Point within, even situations that seemed unfair suddenly reveal new possibilities. If you adhere to the zero laws, reality can bring astonishing transformations. That's the mystery of the universe.

Like the Statue of Justice, the universe only judges whether you maintained the Zero Point or not, nothing else. Human reasoning may find this unfair or unjustified, but everything in the universe happens because it's due to happen. The laws of the universe are executed without a single error.

To harness the power of the universe, one must first understand the heart of the universe.

Stop Feeling

When facing any object, we automatically receive certain feelings influenced by our biases. There are three categories:

1. Pleasant feelings
2. Unpleasant feelings
3. Neutral or indifferent feelings

When we feel good about something, we tend to want to possess or be near it. The reason we desire or want to own something is that we feel good about it. However, it's crucial to note that there's often no verification of whether "it is truly necessary for me." Consequently, we react to positive feelings and chase after things we don't genuinely want, wasting time.

On the other hand, unpleasant feelings instinctively make us want to avoid the source. If escape is impossible, feelings of anger, hatred, and aggression may arise. Yet, these emotions are often subjective, skewed reactions unrelated to the situation, leading to various problems in life.

A neutral feeling, also known as indifference, is often misconstrued as being in the middle, but it's not the Zero Point. It's a state of indecision, where one is too confused to make a choice. Knowing and not choosing is different from not knowing and hence not choosing. This ignorance and hesitation can exacerbate situations or make one vulnerable to manipulation. As soon as our mind leans towards one side due to subjective schemas, it creates various emotions.

To avoid being swayed by emotions, we must first halt the three fundamental feelings - pleasure, displeasure, and indifference. I call this the Law of Non-Emotion. This law means transcending the polarity of likes and dislikes and maintaining the middle ground.

Desires arising from these three feelings are not our foundational desires but superficially imposed ones. Automatic responses to desires are not what we truly wish to achieve. To truly create reality through the zero system, we need to step out of these feelings and find the Zero Point within the state of non-emotion.

Before creating reality through the zero system, it is essential to pause, reassess our long-held desires, and reset ourselves to a state of non-desire. This reset helps us achieve a tranquil state, free from the emotions we've been harboring. We should not hastily jump into creation. First, let go of all your desires and find tranquility. Your heart should always maintain a state of non-emotion. Do you understand?

Zero Law 5 – The Law of Time

The Zero Point is always in the present.

The Secret of the Present

"Carpe Diem! Seize the day!" You've probably heard this often. All enlightened sages emphasize "here and now" because it's the Zero Point where everything can be created.

The Zero Point is located at the exact center of our space and time. The center of space is exactly where I am, and the center of time is the present, situated between the past and the future. Time exists only as a relative concept. The past and the future, though familiar to us, are actually illusions and do not truly exist. Finding the Zero Point means finding the present moment where I am.

How can we find this present moment? On the timeline, the present is merely a brief instant. The future constantly becomes the present and then quickly turns into the past. And so we struggle to understand the true meaning of the present.

To stay at the Zero Point, we must maintain equanimity and not be swayed by emotions. Our emotions are remnants of past experiences and memories stored in our subconscious. They don't exist in the present and are mere illusions. Yet, we often project these past emotions onto our present experiences.

Emotions prevent us from fully being in the present, anchoring us in the past. For instance, finding a ring given by a deceased mother while cleaning might evoke certain feelings. These emotions transport us back to past memories. All emotions work in this way, originating from past memories.

When an emotion arises, we need to recognize its source and its current relevance. Realizing that the emotion is just information stored in the subconscious can help us detach from it and return to the present. This understanding helps prevent errors created by the illusion of the past.

Similarly, future expectations or anxieties are also illusory. Anticipations and hopes for an unmanifested future are mere mirages. They are as meaningless as a mirage to a thirsty person if they can't provide immediate relief. Instead, they may obscure the immediate reality or exacerbate psychological thirst.

The past and future don't exist. If we feel they do, it's only because we're living within the memory of the past or expectation of the future. Getting fixated on the past or future disrupts our balance, leaning us towards one side. Therefore, we should avoid dwelling on any memories or expectations. This allows us to remain fully in the eternal present.

The Zero Point, the present moment, is the eternal instant where only pure awareness exists. It's a state where we accept pure information every moment. At this point, we truly encounter the essence of all things. This is what intuition means—seeing things as they are, unobstructed by our thoughts.

Although it may seem inefficient or unnecessary to constantly recognize reality anew, doing so allows us to stay in a state with zero errors—a state of perfect awareness. Automatic thinking distorts reality, trapping us in delusions.

In other words, the Zero Point is a midpoint where we can see a mountain as a mountain and water as water, without comparing them to other mountains or waters, but merely acknowledging them for what they are.

Maximize your five senses, excluding thought information, to fully embrace pure information at each moment. This will connect you to the "space in-between" in the universe through the Zero Point. You'll experience the mysteries of the universe by encountering the true nature of all things. And this can only happen in the present, right here, right now!

The Magic of the Present

People live within the information of the past. All the information, experiences, schemas, and automatic thoughts stored in our unconscious are from the past. Our lives are a series of moments where the future becomes the present and then quickly turns into the past. What was a future just a second ago passes through the present and becomes the past in an instant. In a space governed by time, nothing remains unchanged; everything rapidly becomes a thing of the past.

No one can predict exactly what will happen in the future. Even the most skilled prophets have only a 50 percent chance of being correct—the prediction will either be right or wrong. The future is shaped by the present, changing as you decide and act. Sometimes, changes can be so vast that they surpass imagination because no one can predict what conditions will emerge in the future.

We can only create conditions for future events. We don't know what additional elements will combine with those conditions to realize the events. Therefore, setting limits on the future based on the present is foolish. The possibilities for the future are endless, depending on what choices and actions you make now.

The past, too, is not immutable just because it has passed. A person doesn't have to carry the disgrace of a careless act from childhood forever. We can change even our past. Of course, past actions have occurred, but we can change our memories of those facts. The past can be a significant wound or a light-hearted incident, depending on the emotions included in the memory. Isn't it astonishing that the past can be remembered more beautifully depending on your current mindset?

This is why living in the present is so important. The basic premise for creating reality according to one's wishes is to remain entirely in the present. However, instead of living in the "magical present" capable of infinite creation, we live in the past that has gone or the future that has not yet come. Salarymen who forget the joy and rest of the present in preparation for the future are living in the future, while singles who cannot find new love because they are stuck in past sorrows are living in the past.

If we could only stay in the present moment, worrying about the past or preparing for the future would become unnecessary. The present holds the key to a new creation that transcends the memories of the past and the expectations of the future. This doesn't mean to live without thoughts or to ignore all your experiences.

It means freeing yourself from memories that limit your possibilities. Don't waste time preparing for the future. Happiness exists only in the present. The only time you can experience is always the present. Try meeting new people every moment, living in new environments, and making new judgments. Eventually, even if you meet the same person or are in the same environment, you'll be able to view them from a fresh perspective!

Do you think you know your long-time partner well? Most likely, it's a misconception. Even if you have shared numerous memories with your partner, it's not easy to see them as they truly are. Your partner is constantly changing, and so are you. And it's not just about people. Every cell in our body is also constantly going through cycles of death and birth. Moment by moment, you are encountering entirely new things!

We should pay attention to the pure information revealed at this moment. Then we can look at each other with more curiosity and marvel at the constantly changing new appearances. Emphasizing living in the here and now, the present, is never too much. It's an essential condition for creating the life you want, transcending the past and future.

Present not only means "gift," but also signifies "being here now." When we exist wholly in the present, we receive the great gift of reality creation from the universe. And this gift of the universe is none other than existing within our consciousness. Bringing our consciousness entirely into the present, that is, originating from the Zero Point.

The Secret of Present-Tense Affirmations

The present is the only place where the power of the universe can be wielded like magic. If you've read self-help books, you already know there are several rules to follow when making affirmations to achieve your dreams.

First, you must use the first-person subject. By using "I am," we declare ourselves as the creators of our reality to the universe. The first-person subject is like a declaration that imprints one's existence in the universe and signals the start of creation. As I often say, you are the most important entity in this world. The world exists for you!

The tense of the affirmation must always be in the present. The past and the future do not exist. The place where magic happens is in the present.

An example of a correct affirmation for reality creation is:

"I am comfortable, prosperous, and have a lot of money."
(First-person subject + present tense)

Our desires often arise from a sense of lack in the present. Therefore, without realizing it, we tend to express affirmations in the future tense. Phrases like "I want" or "I wish" are all in the future tense. Hoping for creation from a state beyond the Zero Point is an unfeasible error.

Examples of incorrect future tense affirmations
"I will pass the job interview."
"I want to become comfortable and prosperous."
(First-person subject + Future tense)

When making future tense affirmations, our consciousness is not in the present, the scene of creation, but in the imagined future. That's why, no matter how much we wish, it won't come true. Future tense affirmations remain forever as future tasks, never becoming proper spells of creation. Remember, the point of creation begins at Zero Point, and the affirmation of creation must always be in the present tense.

Sometimes, we fear imagining a future that seems realistically impossible. For example, someone who has been overweight their entire life, with no success in dieting and a history of being mocked, might not even consider attempting a new approach to achieve a slimmer future. The anxiety arising from the subconscious can hinder the creation of present tense affirmations. Even if one creates a wonderful present tense affirmation, they might not fully believe in it and remain trapped in the negative feelings of the past. In a state filled with negative energy, it's hard to shake off the anxiety, even with present tense affirmations.

In such cases, a good method is to use self-acceptance affirmations that acknowledge and embrace this anxiety. Completing a positive present tense self-acceptance affirmation like, **"I am [lacking state], yet I fully accept myself,"** can help you move beyond internal resistance and progress toward creating the desired future.

Self-acceptance affirmations are widely used in the Emotional Freedom Technique (EFT). EFT effectively utilizes self-acceptance affirmations to overcome the feeling of lack in the present.

"I am currently in pain, but I fully accept myself."
(First-person subject + Present tense self-acceptance affirmation)

Self-acceptance affirmations have another benefit: they prevent inadvertently creating affirmations with negative expressions. For example, if someone repeatedly affirms, "I am not in pain," to alleviate physical discomfort, it's not a positive affirmation for creating reality but rather a denial of reality, inadvertently reinforcing the denied reality. However, using the positive self-acceptance affirmation of EFT naturally transforms the negative perception of the current situation into positive energy.

The rules for affirmations for reality creation are as follows:

1. **Declare with the first-person subject.**
2. **Use the present tense.**
3. **Use positive expressions.**
4. **If the current situation is too hard to accept, embrace reality with self-acceptance affirmations.**

By following these four rules, we can always make requests to the universe from the Zero Point for the reality we desire. From now on, consistently check if you're correctly making requests to the universe and enjoy the process of your life changing!

Creating Personal Affirmations for Reality Creation

1. Write down your current situation and what you desire. Example: Currently, I lack money. I wish I had more money. I don't have a partner. I wish I had a wonderful partner.

2. Transform the content from step one into present-tense affirmations in a single sentence. Example: "I am free from lack."

3. Check if the affirmation is not negative. Compare it with reality and ensure it's a positive present-tense command. If it's correct, keep it as is; if not, modify it. Example: "I am free from lack." → "I am prosperous."

4. The affirmation from step three is your magical spell for creating the desired reality. Place it somewhere visible daily, and merely acknowledge it without effort or intent. Simultaneously, be mindful of the other zero laws, and work towards creating conditions that enable the realization of your affirmations.

Zero Law 6 – The Law of Emptiness

The more I empty myself, the closer I get to the Zero Point.

State of Zero Existence

The sixth Zero Law is about "emptying." Zero (0) signifies nothingness. To achieve a state of nothingness, you must let go of everything you have perceived as "self." This is the ultimate state of consciousness emphasized by enlightened individuals throughout history.

What does the state of "no self" mean?

Water and oil are liquids of completely different densities. They appear similar but are so distinct in nature that they remain separate. The only way they can become one is by adding an emulsifier, which removes the separation between them. The emulsifier allows the oil, which does not mix with water, to disperse evenly in it. Meeting the emulsifier, oil loses its original properties and starts blending into the water, leading to a new substance that is neither just water nor oil.

The various problems we encounter in life mostly arise from a lack of understanding between ourselves and others. Like water and oil, we are separate entities. We can never fully understand or become one with another. However, just like the emulsifier, there is a way to dissolve "self" and merge with others. We often expect others to change while remaining unchanged ourselves. But it's challenging to change someone else's heart. To become one with another, the only way is to dissolve oneself.

When you blend into others, the separate "I" no longer exists. "Others" cease to exist too. Only "we" remain. In the moment I disappear, I am not gone; rather, I become everything. All the problems that troubled us vanish, and the reality we longed for begins to materialize.

This is the special magic of Zero (0)—connected to infinity. By letting go of my desires, thoughts, and everything I own, I become Zero. At that moment, we connect to the universe's infinite energy field, the Zero Point Field.

The Zero Point Field

We often assume that empty space in the universe contains nothing. However, physicists have discovered a fascinating fact: even the vacuum at absolute zero temperature (-459.67°F) is not an empty Zero State but is filled with something. In a vacuum, where theoretically nothing should exist, there are still pairs of particles and antiparticles appearing and vanishing in incredibly brief moments. Even at the theoretical absolute zero temperature, where particles should be "frozen" and immobile, there is still a presence of infinitesimally rapid movements. This is what constitutes the Zero Point Field (ZPF).

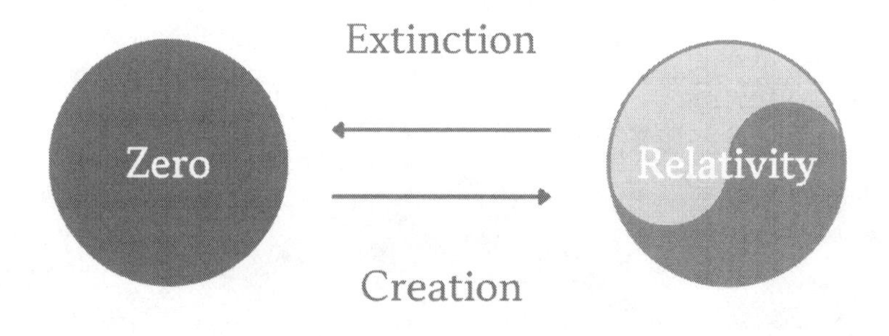

Creation and Dissolution in the Universe

1. All relativity returns to zero (dissolution).
2. From zero, everything splits into relativity (creation).

According to Newton's laws of motion, for an object to remain still and not move, the forces acting upon it must cancel each other out to zero. Like in tug-of-war, when equal forces are applied from both sides, the rope does not move. Although it appears still, it's actually engaged in a subtle struggle. This principle applies universally. Even though macroscopically it seems still, at the microscopic level, it's constantly vibrating near zero. This is the immense power of the Zero Point Field.

The reason the Zero State is a space of infinite potential is that the universe exists in the form of a quantum vacuum, a densely packed state of various energies in the Zero Point Field. The space of variations in the Zero Point Field is a space where infinite transformations and creations are possible. All attributes exist near zero and manifest into "appearances" when a certain property is bestowed. The combined energy of particle pairs vibrating in the quantum fluctuation in just $1m^3$ of empty space could evaporate all the water on Earth—a testament to its immense energy. Some physicists are exploring ways to harness this infinite free energy of the Zero Point Energy. If successful, it could be a groundbreaking alternative energy source. While scientists are striving to find a physical way to harness this energy, let's leave that to them. We know we can influence the quantum dimension, the Zero Point Field, with our consciousness and so affect the macroscopic dimension of reality. Although it might seem insignificant, the principle is the same.

Physicists like David Bohm and neurophysiologist Karl Pribram turned to holography, the laser-generated 3D images, to explain inexplicable experimental results. They concluded that the material universe exists as a holographic information field at the quantum level. As the poet Friedrich Schiller put it in his *Philosophical Letters*, "The universe is God's grand thought"—this insight might mean that the material universe is reality shaped by divine thought, and within it, humans shape their reality with their thoughts. Our observed particles influence and change each other, forming our world.

Scientists like Barbara Brennan, a former NASA atmospheric physicist, and Edgar Mitchell, an Apollo 14 astronaut, have said that our body's information-energy field resonates and exchanges information and energy with the cosmic energy field. Ancient Indians called the space akasha, where all information of the universe is recorded. The "Book of Heaven," "Spiritual Realm," and such seen by seers and visionaries across cultures are the quantum information (potential reality) recorded in the Zero Point Field.

Erwin Laszlo, a philosopher of science, argued that the Zero Point Field is the film (hologram) that records the universe's information-energy (wave information). Holography technology uses a hologram (film) to record the interference pattern of light reflected from an object, possessing mystical properties like totality, non-locality, and four-dimensionality. Moreover, theoretically, an infinite number of objects can be recorded on a single hologram. The Zero Point Field, as an extremely sensitive medium, records all past, present, and future waves occurring in the universe, projecting the 3D image of the material universe on the "present" stage. The hologram's dimension is a 4D world where past, present, and future coexist simultaneously.

Time doesn't exist the same for everyone. Einstein said time flows differently depending on where and in what situation one is. Your subjective time will start to apply differently as you get closer to cosmic consciousness.

Ancient Indians said the *akasha* records all information of the universe. Similarly, "Heaven's Book," "Spirit Realm," seen by prophets and seers are the waves of past, present, and future reality recorded in the Zero Point Field. The Zero Point Field is where all creation possibilities exist, though appearing empty. It's where the magic of turning wishes into reality happens on the stage of creation.

How can we connect our consciousness to the Zero Point Field? The Zero Point Field, as an empty film, has no attributes—its sole property. To resonate our mental energy with it, our consciousness must also be a Zero State, attribute-less. This is why we need to practice letting go of attributes we believed were "us." By emptying our consciousness to a state of no attributes, we can connect to cosmic consciousness. Then, with a single pure wish, the Zero Point Energy will resonate (like the energy powering a TV), amplifying the signal, and our wish will manifest as a reality on the macroscopic material world screen. This is the law and promise of the universe.

Just like that, we gain the ability to use the infinite energy of the universe to freely realize our dreams as creators of reality!

The Nothingness

Everything continuously arises and vanishes, so nothing is ever permanent. Buddhism calls this impermanence or *anitya*. Existence isn't about some fixed entity; it's the phenomena themselves, constantly emerging anew. A deeper insight reveals that all seemingly fixed phenomena are actually part of a continuous flow of change, with everything in the universe constantly arising, ceasing, and transforming in accordance with cosmic flow.

In Eastern Qi philosophy, physical form is seen as a temporary congregation of Qi (energy), and when Qi disperses, the physical form also ceases to exist. This concept, evolved from the theory of Form and Qi (Xing Qi) in the *Huangdi Neijing (The Yellow Emperor's Classic of Internal Medicine)*, is remarkable in that people five thousand years ago were discussing ideas similar to those of twentieth-century modern scientists.

Molecular biology, based on the premise that humans are made of matter, explains the human organism as a collection of cells composed of molecules and atoms. Sherwin B. Nuland, a professor at the Yale University School of Medicine, said, "The genes that determine our essence are just molecules of protein, composed of atoms that have existed for 3.6 billion years. What was needed for life to arise from them was energy."

However, from the quantum perspective, there's no "body" as such. It's just a collection of particles flickering in and out of existence in vast empty space. Moreover, since the cells in our body are also constantly being created and destroyed, there's nothing permanent anywhere. Even our own bodies are embodiments of impermanence.

The Nonexistence of Self

Since everything in the universe is impermanent, the notion of "I" or "self" extracted from these transient phenomena is merely an illusion. Buddhism emphasizes this through the concept of *anatta* (no-self), asserting that there's no fixed entity that we can identify as "I."

The self is less a static state and more a collection of changeable attributes. No-self (anatta) refers to a state devoid of any attributes. We often mistakenly identify ourselves with the roles we play in various situations. A homemaker might identify as a mother of a child, an employee as a worker at a company. At a department store, one becomes a customer; with a partner, a lover. However, these identities are just relative attributes defined by circumstances, lacking any permanent "self."

Then, where is the true self, and how can one encounter it? To meet the true self, we must begin by observing the attributes we've always considered as "I." Observe yourself breathing, thinking, reacting, sleeping, and waking up.

In this process, you'll realize a new "self" observing these various attributes. The self that adapts and reacts to situations through thoughts and actions is not a fixed entity but a series of transient states. The true nature of our existence is in the unchanging observer, quietly witnessing these changes. This enduring observer, rather than the fluctuating self, represents our true essence.

Observer

So who is this "Observer?"

In Hinduism, the immutable true self is referred to as *Ātman*. While all things are constantly changing and so lack a fixed "self," the "Observer" or the "state of observing consciousness" definitely exists. This is the true "self" and the originator of creation.

To connect with the energy of creation, it is necessary to awaken this Observer within. The Observer is not some other being, but the true self encountered at the Zero Point, hidden beneath the false self.

Observing all subtle changes within and external phenomena, we encounter another "self" that observes everything. This Observer exists as consciousness without any attributes or form. When our state of consciousness transcends the smaller "self" and extends into a subtle, universal dimension, our intentional observation can bring about changes in the universe.

This state of consciousness is what connects to the Zero Point Field. Just as the power energy of a TV receiver amplifies weak signals to produce clear images, the state of consciousness of the Observer can amplify desires to create vivid realities. This is the state of consciousness of a creator who can utilize zero-point energy.

However, the Observer does not respond to unbalanced emotions, deficiencies, and desires mixed with falsehoods. The Observer transcends all experiences and emotions, maintaining perfect balance in a state of serene equilibrium.

In Buddhism, *prajñā* (wisdom) does not refer to specific content. Instead, it signifies a holistic perspective. Ignorance is viewing things from a biased perspective, while wisdom is seeing from all angles. As the true "self" or the Observer, we can create our reality with prajna and zero-point energy. Pure desires become acts of creation in themselves. We are not separate from the genie of the lamp, so there's no need to wish upon the genie anymore.

To reach the state of consciousness of the Observer, we need to understand and practice the principles of zero law. Only then can we encounter the mysteries of the universe. Remember, this world is indeed a realm of infinite possibilities!

Practice Becoming the Observer

To exist as the Observer, it's essential to regularly practice recognizing oneself as the Observer. Let's try to detach ourselves from all the attributes we believed to be "us." We have been living a passive life, preoccupied with adapting to given situations. This practice will transform our perspective into an "active" one and help us remain in the state of the Observer's consciousness.

1. Call your own name as if you are calling someone else.

2. Speak your name and your current situation in the following sentence format. If a new feeling arises, change the sentence and repeat.
   Example: John Doe is currently looking at _____. /
   John Doe currently desires _____.

3. By calling our own name as if it's someone else's and objectively describing our actions, we become the observing subject. Then, we start to discover the issues with "me" and begin to observe every aspect of life accurately. As many things we believed to be "us" fall away, we find ourselves existing in a new world—essentially, remaining at the Zero Point.

The Mystery of Letting Go

The realization of impermanence and no-self is not exclusive to Buddhism. In Christianity, there's an emphasis on "laying oneself down completely before God" and crying out, "Let Your will be done." Yong-Gyu Lee, the author of the bestseller *Letting Go,* which explains how to live by listening to God's voice, mentioned the reason God asks us to let go is that "only when we let go, does it truly become ours."

When we let go of being "I," we encounter our essence, the "true self." Religious people often say, "When I let go of myself, I met God, I had a mystical experience." Letting go of "I" means to release all the notions that have been defined as "I." This allows us to experience mystical events beyond ordinary capabilities.

Religious people might describe these phenomena as "God's will," but it might be more accurate to call it the "law of the universe." This is because the creation of reality is a universal miracle that can be experienced by all humans, regardless of religion. People who embark on finding their true selves within the universe and awaken their true nature ultimately encounter the "it" that all religions speak of. All religions exist for a single purpose, which is the complete dissolution of the "false self."

The dissolution of the "false self," becoming one with everything, is the path to entering the state of Zero, achieving a miraculous harmony.

Discard yourself. Empty yourself. And become still.
There is no "I."
Only "it" exists, quietly, forever.

Zero Law 7 – The Law of Zero

Stopping is another beginning.

Altered States of Consciousness

Professor Charles Tart of the University of California coined the term "Altered States of Consciousness" to describe non-ordinary states of consciousness. This refers to the state of consciousness entering a Zero State through spiritual practices like Zen meditation or contemplation. In this state, we surpass everyday limitations and exhibit astonishing abilities. Sometimes, people display seemingly impossible, superhuman capabilities; this is the power of the altered state of consciousness.

For instance, to save their children in danger, mothers have been known to run incredibly fast or lift cars with tremendous strength! How do these powers suddenly emerge? By momentarily forgetting their self-imposed limitations and focusing solely on the singular intention to save their child. In that frantic moment, the mother becomes a Wonder Woman.

That moment of, "My child—no!" is an altered state of consciousness. The state of consciousness during shamanistic rituals or when a medium communicates with the spirit of the deceased is similar. In this altered state, all human limitations vanish. We connect to infinite energy and start displaying abilities beyond imagination.

However, there's a condition to enter this altered state of consciousness: one must infinitely zero out their consciousness. Zero refers to a state of clear, awakened consciousness with no thoughts. The reason practitioners meditate is precisely to practice cutting off thoughts.

Creation of the New When Thought Ceases

Ordinary people often feel uncomfortable with the state of not thinking, simply being. This discomfort stems from being brainwashed into believing that constantly planning and acting is a necessity for survival.

As a result, people always rack their brains looking for solutions to problems. Some even resort to brainstorming, randomly spewing out ideas from their minds, in the hope of finding new insights. Although iterative thinking can lead to great ideas and help in exploring various possibilities, it's more about squeezing ideas out, rather than truly creating them.

Brainstorming is merely laying out what we already know in a random manner. It doesn't bring in unknown information. For more creative ideas, one needs to look outside, not inside. Most creative ideas are born from inspirations that we never imagined.

Genius artists create new works from inspiration. Michelangelo, who sculpted the David, said, "I saw an angel in the marble and carved until I set him free." People believe creative individuals meticulously plan and conceptualize their creations, but in reality, creative people often say they just "felt it one day," without attributing any specific purpose or reason. They follow inspiration.

Groundbreaking ideas often pop up in the bathroom. While it might sound odd, if you think deeply, the bathroom is a place of complete relaxation.

Precisely when people let go of the notion that they must do something creative ideas start to emerge. Google, a global IT giant, has designed its office like an amusement park, where employees can rest, enjoy, and work anytime. This approach challenges the traditional management style of "squeeze your brain at the desk."

This approach proved successful. Google has established itself as a company that developed the world's top search engine and numerous globally used programs, receiving high praise for its current performance and future growth potential.

"This was my everyday life. I was alone. All I needed was a cup of tea, light, and music. That was all I had," said Steve Jobs, the co-founder and former of Apple. Jobs always preferred being in a quiet space alone. His extraordinary insight and creativity were rooted in meditation practices.

Carl Jung also said, "People called geniuses can communicate well with the cosmic consciousness and thus gain various excellent abilities."

When thoughts cease and we relax, our brains enter a comfortable alpha wave state. As we relax deeper in the alpha state, we progress to the theta wave state. In this theta state, we unconsciously issue commands to the universe. The waves of consciousness emanating in the theta state resonate with the answers we are genuinely seeking in the hologram of the zero point field, the universal library, and make us recognize them as ideas. That's when we exclaim, "Yes, this is it!"

Step away from the compulsion to do something, the need to achieve something. Give your body and mind a rest! When you turn off worries and thoughts and stay at the Zero Point, you will encounter a new world.

Earth's Frequency of 7.8Hz

German physicist Winfried Otto Schumann discovered that the Earth's inherent frequency is 7.8Hz, leading to the establishment of the Schumann Resonance Theory. It's believed that human brainwaves can resonate with the Earth's natural frequency. In fact, the Earth's inherent frequency of 7.8Hz corresponds to the theta wave range in brainwaves.

Theta waves represent a state of semi-awake, semi-sleep, or hypnagogic state. In this state, the secretion of brain hormones is activated, enabling experiences of tranquility, happiness, and sometimes even transcendental psychic abilities.

However, electromagnetic waves emitted by human-made electronic devices disrupt our brainwaves, causing them to resonate with frequencies other than the Earth's. The artificial vibrations generated by machines are not beneficial for our bodies. This principle is reflected in the old saying, "Returning to nature cures diseases." By realigning our frequency with that of the Earth, our ancestors aimed to activate the immune system, embedding this wisdom in their practices.

When in an altered state of consciousness, our brainwaves stay in the theta state. Theta waves (4-7.9Hz) are a slightly lower frequency range than the alpha waves (8-13.9Hz), associated with meditation and the cessation of thoughts. When the alpha and theta waves resonate together, around the mystical point of 7-8Hz, the right hemisphere of the brain becomes rapidly active. At this point, our bodies resonate with the Earth's natural frequency, allowing us to fully receive its energy.

The universal frequency (7.5Hz) is slightly lower than the Earth's frequency (7.8Hz), also in the theta wave state. Initially aligning with the Earth's frequency, and then lowering brainwaves further to resonate with the universal frequency, allows us to freely harness the universe's energy.

The universe exists as an infinite quantum vacuum, a space of possibilities. By lowering our consciousness to the theta state and directly inputting our desires into the zero point field, the infinite energy of creation can be activated. The universal frequency is, in essence, the frequency of the zero point field, a repository of infinite energy. Through practice in controlling consciousness, we can intentionally lower our brainwaves to resonate with the universe.

The singing bowl, as an addendum, is a tool that generates vibrations. Its slow vibrations help synchronize our brainwaves with the theta state, facilitating deep relaxation. With the aid of a singing bowl, anyone can easily lower their brainwaves and enter a state of consciousness resonating with the universe. This is the true meaning of singing bowl meditation—a seemingly magical tool that enables easy access to a Zero State through vibration.

The reason we often experience incomplete creation is that our current consciousness only occasionally and accidentally resonates with the universe, with most of our time spent in the beta wave state of everyday consciousness. To increase the success rate of reality creation, practicing to maintain a Zero State of consciousness that resonates with the universe is essential.

Zero Law 8 – The Law of Doing Nothing

Create everything by doing nothing.

The Principle of Healing

Sometimes, we encounter individuals with the mysterious ability to heal people. They often use prayer or energy transfer through their hands. When asked how they heal, they usually say, "I just received the energy from the universe and transferred it." In healing practices like Reiki or Pranic Healing, the healer merely acts as a channel to pass the universal energy to the patient. If the healer's personal intentions interfere, the energy won't transfer correctly. Therefore, healers must abandon their intentions and desires, simply facilitating the proper transfer of universal energy. The key to healing lies not in the healer's energy but in how much of the universe's energy is transmitted.

> All processes of divine healing presuppose cooperation with the Holy Spirit. A healer is just a mediator of the universe's plan. When praying, we are essentially seeking the life-giving ether. The prayer of a healer should not be mere words but a genuine resolve and action to help a suffering neighbor. A healer who genuinely desires to help will find their life force replenished, and there are no known failures in healing with such true aspiration.
>
> —From *The Saint of the Mediterranean Daskalos*

This principle of healing also applies to the creation of reality. To use the energy of the universe, we must align our state of being with the Zero Point, where no emotions, thoughts, or desires exist. Only the true "I," the "I" that disappears, remains. Just as the healer is left with only altruistic love for the patient's healing, we too must adopt this state. In fact, healers have nothing to do. They mustn't think or do anything. When healers empty their minds, the universe uses them as a conduit to flow into the patient. Then, miraculous healing occurs as the universe's energy enters the patient's body, awakening their natural healing abilities.

Marilyn Ferguson in *The Aquarian Conspiracy* mentions that the human body's tissues and organs possess natural healing powers, which she refers to as the "perfectly intact doctor" within or the "wisest doctor in the universe." If we can become healers ourselves and awaken our natural healing powers, we will live a life free from illness.

In religious healing cases, this principle is also evident. Religious healers completely surrender to God in a state of total emptiness and seek healing through the power of the Holy Spirit. Miracles naturally occur independent of their intentions.

I have discovered that this principle also operates in singing bowl healing. The vibration of the singing bowl, too, awakens a person's unique energy when the healer is in a complete Zero State. The healing is actually brought about by our own bodies. Therefore, I always teach healers to empty their minds and just serve as channels for healing. The more the healer empties their mind, the higher the efficacy of healing. Any intervention of the healer's intention halts the healing process. Even the magical vibration of the singing bowl, which can bring our consciousness to the theta state, is somewhat transformed by human consciousness energy. Ultimately, our use of consciousness energy is most crucial. Our subtle consciousness energy interacts with the vibration, determining the success or failure of healing. Even with a mighty singing bowl, proper healing can't occur if one is not ready to use zero energy.

Action (Youwei) vs. Inaction (Wu Wei)

Youwei refers to the endeavor to achieve desired outcomes through willpower and physical actions, ignoring the universe's zero system. All human-made civilizations are creations of Youwei, inevitably influenced by the complex interplay of physical environmental causality. Our world is shaped by the wills and desires of countless individuals. If what you want conflicts with another's desires, imagine the amount of obstructive energy that can arise.

In contrast, *Wu wei* minimizes the influence of physical causality and moves energy invisibly without depending on the physical environment. Wu wei means, "Doing nothing" or "without intent." It might seem contradictory to effect change through Wu wei, but that's just a rational human question. The universe is inherently paradoxical, making Wu wei, in reality, more powerful and fundamental.

Our world of phenomena is governed by Youwei. The phenomenal world, as explained earlier, has no substance on its own and exists in conjunction with the non-phenomenal world. The non-phenomenal world, being unmaterialized, is invisible but serves as the source behind the materialized aspects. Wu wei operates on the principles of the non-phenomenal world, and since the phenomenal world is created from the non-phenomenal, Youwei is ultimately subject to the laws of Wu wei.

> Youwei relies on various causal forces (Karma), but Wu wei is a powerful act that does not depend on causality. Like how a less confident person tends to rely on others, but a confident individual is independent. Since Youwei is born out of causality, it lacks inherent nature. Focusing solely on causal forces leads only to suffering. It's like a treasure, similar to a hot lump of gold—if you hold it in your hand, you get burned.
>
> —From *Abhidharmakosabhasya*

To connect with the non-phenomenal world and move the phenomenal world, we must adjust our mental energy. This requires understanding the zero system—the language of the non-phenomenal world that programs the phenomenal world. When we use the energy of Wu wei, what changes occur in the phenomenal world? Is Wu wei truly more effective for creating reality than Youwei? To find answers, it's necessary to compare the effects of visualization techniques and mindfulness meditation.

Visualization vs. Mindfulness

The technique of visualization, where you vividly picture and repeatedly reinforce your desires in your mind, is an effective method for creation. However, it essentially involves active effort to bring about change, making it a form of "doing" or Youwei. To compare the effects of Youwei versus Wu wei, I planned an experiment.

In 2010, I wrote a thesis titled "Study for Chronic Pain Reduction Effect of Mindfulness Meditation by Doing Nothing—Focus on Transpersonal Psychiatry." It focused on an experiment to determine which was more effective in alleviating chronic pain: Youwei, represented by visualization techniques, or Wu wei, represented by mindfulness (Vipassana) meditation.

John Kabat-Zinn, founder of the Stress Reduction Clinic at the University of Massachusetts Medical School, inspired by *Vipassana* meditation, developed the MBSR (Mindfulness-Based Stress Reduction) program to aid in pain reduction. This program has significantly contributed to reducing symptoms and suffering in a wide range of conditions, from chronic pain to psoriasis, and has been adopted in over two hundred health-related centers in the United States alone.

The key healing mechanism of meditation lies in Wu wei—simply observing and being aware of what's happening without any specific intent or action. How can this simple method of just observing bodily sensations impact pain treatment?

I gathered fifteen students in their twenties and thirties suffering from chronic pain through the internet. They were divided into three groups of five. The first group practiced mindfulness meditation, the second practiced concentration meditation focusing on visualization, and the third did nothing. Over five weeks, I measured stress, depression, and pain levels in each group. The participants were beginners or had little experience with meditation.

For the mindfulness group, I instructed them to follow John Kabat-Zinn's MBSR program, which included walking meditation, sitting meditation, and body scans for fourty minutes each week.

In contrast, the visualization group was guided to meditate with a specific intention. They focused on their breath, saying "I am" while inhaling and "healthy" while exhaling. For fourty minutes, they repeated these affirmations internally while visualizing themselves in improved health.

After five weeks, what were the results?

Five-Week Mindfulness Meditation Program

1st Week	Orientation 10 mins, Sitting Meditation 10 mins, Body Scan 20 mins
2nd Week	Sitting Meditation (Observing Breath) 20 mins, Body Scan 20 mins
3rd Week	Sitting Meditation (Observing Breath) 20 mins, Body Scan 20 mins
4th Week	Walking Meditation 20 mins, Body Scan 20 mins
5th Week	Walking Meditation 20 mins, Body Scan 20 mins

As shown in the chart, the group that practiced mindfulness meditation experienced significantly greater pain reduction than the group that practiced visualization. Additionally, in verbal reports, participants in the mindfulness group actively stated that their pain had disappeared, and some could even pinpoint when it happened. Conversely, the visualization group only reported feeling somewhat lighter in body and mind.

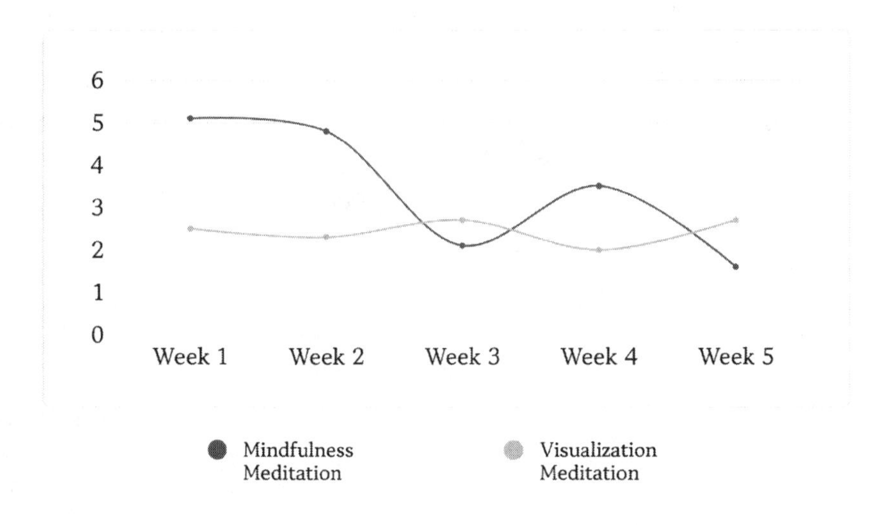

Despite only meditating for forty minutes a day over five weeks, the mindfulness group demonstrated exceptional pain reduction. In contrast, those who visualized affirmations like "I am healthy" during the same period did not exhibit notable changes. Why this discrepancy?

Wu Wei (Non-Action)	Youwei (Action)
No intention.	Exerting a specific intention.
See the object as it is and do not distort it.	Distortion occurs when the object is viewed with thoughts in mind.
Energy is concentrated on the object (essence). → The object disappears.	Energy is concentrated on the image (not the essence). → The object is transformed.
Results: Accuracy of concentration increases because energy is concentrated on the target.	Results: Accuracy in concentration is reduced because energy is applied to the thought rather than the target.

When we observe without any intention (thought), we confront the subject's essence without distortion. Mindfulness meditation allows us to fully perceive the subject, realizing its impermanence and essence, and ultimately leading to its dissolution. In contrast, visualization involves creating an image based on one's desires and focusing energy on it, ignoring the subject's essence and indulging in one's imagination. The energy concentrated on the visualization does not transform the essence but rather attracts something else that resonates with that image, leading to less tangible effects.

In summary, mindfulness meditation corresponds to Wu wei, while concentrated visualization corresponds to Youwei. The underlying mechanisms of Wu wei and Youwei differ, which affecting their efficacy. Mindfulness, by observing things as they are, operates under the principle of Wu wei. Understanding that the body, like pain, has no real substance, the physical discomfort disappears. This is the essence of the Zero State, the void.

"The myriad things of the world are born from being, and being is born from non-being."
—Laozi, *Tao Te Ching*, Chapter 40.

Body Scan Meditation

If you're currently experiencing pain in your body, try a body scan.

The body scan is a practice similar to how a scanner operates, methodically moving your attention through your body from head to toe, simply observing what is there. While performing a body scan, you might notice at some point that the pain dissipates.

1. Sit comfortably in an upright position in a quiet space.

2. Take slow, deep breaths in and out.

3. Expand your awareness to encompass your entire body. Then, like a scanner scanning the body, slowly shift your attention from the top of your head to your toes, locating any areas of pain.

4. When you find a painful area, simply acknowledge it: "There is pain in my shoulder," or "There is pain in my knee." It's important not to judge the pain as "severe," "needing to go away," or "unpleasant." Just recognize that pain exists in that location.

5. Once you've scanned down to your toes, start again from your head and repeat the process. If you find areas of pain, acknowledge their presence and continue scanning.

6. After about twenty minutes, gently bring your breathing back to normal and gradually expand your consciousness from your body to the external world, returning to your usual state of mind. You can extend the duration gradually as you get more comfortable with the practice.

If, during the body scan, you find your attention drifting to other thoughts (like wondering when the pain will go away, whether this will work, or focusing on the discomfort), simply notice this and gently redirect your focus back to scanning your body. Remember, just acknowledge the presence of pain without trying to do anything about it.

Life of Zero, Wu Wei Ziran

Wu wei ziran, as described by Laozi, refers to the myriad things in existence that exist and operate on their own, without human intervention. This concept emphasizes that nature follows the laws of the universe, creating and dissolving in perfect harmony. Everything in *ziran* (nature), including our genetic makeup, inherently knows how to survive and thrive without being directed.

This self-operating system is also present in our bodies, seen in the autonomic immune and self-healing systems that have sustained human survival long before the development of medical science. For example, a newborn deer instinctively knows how to walk and recognize its mother. Everything in nature inherently knows how to live—this is the essence of Wu Wei Ziran, signifying a universe where life and environment are in perfect harmony without human intention.

We are born with the ability to communicate with the universe. Ancient humans knew how to live in harmony with nature, but as civilizations developed and settled, humans began to prioritize human-centric thoughts and learned artificial ways of existence, focusing on efficiency and rationality. However, nature doesn't operate on efficiency or rationality; it is already in a state of perfection.

It's time to rediscover the order of the universe. Nature, operating flawlessly on its own, does not require human meddling. As part of nature, our role is to coexist, following its laws. Living a life of zero (Wu Wei) means letting go of all artificialities. It's not about rejecting civilization and retreating into the wilderness; it's about an attitude shift. Embracing our natural selves and stepping away from self-centric thinking leads us to harmonious coexistence. Like a symphony, where each instrument doesn't assert its sound but harmonizes with the whole, a beautiful collective emerges.

Tune into the rhythm of the universe.

By letting go of selfish desires and maintaining our minds at Zero Point, we connect with the infinite energy of the universe. Our lives then flow naturally, allowing the inherent potential within us to flourish.

We possess everything needed for a rich life, including the best personal assistant (the universe) and a perfect command system to manage all aspects of life. We just need to trust and utilize these resources, as our task is essentially nothing.

This is the essence of Wu Wei Ziran—by "doing nothing," we unleash our infinite abilities.

Chapter 4. Order

The Process of Creating Reality Within

Something mysteriously formed,
Has always existed
Silence and void,
Solitary and stable,
Circulating and ceaseless.
It may be mother to ten thousand things.
I do not know its name,
So call it Tao.
And arbitrarily call it great.

Great, therefore it flows onward.
Flowing far.
From far, returns.

Tao is great,
Heaven is great,
Earth is great,
Humanity is also great.
These are the four great powers of the universe,
And the humanity is one of them.

Humanity follows the earth.
Earth follows heaven.
Heaven follows Tao.
Tao follows what is natural.

— Tao Te Ching, Chapter 25

Creating Reality

We have the power to create anything we want, and the method of creating reality is simpler than we think. Let's recap.

Our thoughts carry unique frequencies, and the more intense the thought, the stronger the frequency. This frequency spreads through the medium of the universe, resonating and attracting energies and materials compatible with our creation into our reality. These may manifest as unseen forces, physical matter, or new relationships.

As our environment changes to align with our desires, the conditions for realizing our dreams swiftly fall into place, ultimately materializing in our reality.

The principle is simple. However, the reason why we often manifest our fears rather than our desires is that the frequencies of thoughts entrenched in our subconscious overpower the frequencies of our aspirations, creating reality on their own. Hence, we must first eliminate the obstructions in our subconscious before focusing on our desires.

How to Create Reality

1. **Deeply Understand What You Truly Desire:** Reflect on what you really want.
2. **Recheck Your Desires:** Make sure these desires aren't influenced by others.
3. **Eliminate Schematics and Information in the Subconscious:** Clear out subconscious patterns and information.
4. **Identify the Cause for Desired Results:** Understand what needs to be in place for your desired outcome.
5. **Align with Material Conditions While Focusing on the Cause:** Think about the cause and arrange the necessary conditions.
6. **Act If Necessary:** Sometimes, action is required to manifest your desires.
7. **Be Patient for Results:** Realize that results take time and continually review the process.

Let me share a real-life example. Choi Eun-jung, who learned the reality creation technique from me, had an interview scheduled with a volunteer organization, coincidentally on the same day as my first public lecture on the Zero System. Although she hadn't received the interview time, she suggested recording an interview on a camcorder in case the times clashed.

I asked her to reconsider.
"Eun-jung, do you have a camcorder at home?"
She said, "No."
So I told to her. "Why not create your reality? It's not Saturday yet, and the interview time isn't set. Align the interview schedule with yours and refocus your mental energy." Realizing she had forgotten to apply the reality creation technique, she decided to try it.

The next afternoon, she joyfully informed me that her interview was set for 6 PM, after the lecture, allowing her to attend both. I commended her and encouraged her to continue actively creating her desired reality. She became much more proactive in manifesting her desired realities afterward.

By continuously being aware and freeing ourselves from the subconscious, we can control our lives and create the reality we want. There are no limits to our future; they are merely the boundaries of our imagination. However, we must manage the karmic consequences that might interfere with these changes. Until we can control these consequences, we might experience incomplete creation, with some things manifesting and others not.

Hidden Variables in Reality Creation

In this universe, no result occurs without a cause.

Even if everyone desires a prosperous and happy life, if they don't create the causes for such a life, it remains an empty dream. It's like hoping for a house to fall from the sky. Despite the popular belief that "if you wish it, it comes true" or "what you think, you attract," without actively creating causes, wishes remain unfulfilled. For example, if you want a good job, you must actively acquire relevant knowledge and submit resumes to create your opportunities. It's not enough to sit at home, idly dreaming of employment.

Classical physics viewed all events in the universe as occurring due to preceding events, based on Pierre Laplace's physical determinism, which suggested that the future could be predicted if the current conditions were precisely known. Buddhism, on the other hand, teaches that a person's intentional actions become the cause, leading to inevitable reactions as results. These intentional acts are called "karma," and the resulting reactions are known as *vipāka*. All phenomena, both mental and material, are interconnected in an endless chain of cause and effect, from which nothing can escape. This concept is known as *Pratītyasamutpāda* or "Dependent Origination."

> "The law of Dependent Origination is not something I created. Nor is it created by any other absolute being. It is a law that always exists in the universe, whether I, the Buddha, appeared in this world or not. I merely realized this law, achieved universal enlightenment, and explained it for all beings. Dependent Origination means "this exists because that exists, and this arises because that arises.'"
>
> —From the *Jātaka-Agama*, Volume 12, Section 299, on Dependent Origination.

To create a result, we must perform certain actions. For instance, to get a job, one needs to look for opportunities and submit resumes. In other words, the beginning and the process become the cause for all the results. This law of cause and effect tightly controls our lives.

As the saying goes, "You reap what you sow." We only obtain results based on our causative actions. There are significant causal relationships that determine the course of our lives, but there are also countless minor ones in everyday life. Even if you plant beans, they won't grow if left unattended. They need appropriate care, like water, sunlight, and fertilizer. Harvesting beans is achieved through the minor causes of nurturing them day by day.

Although the realities unfolding at each moment are created by past causes, that's not the whole story. We also have opportunities to make new choices every moment, which then become new causes for our future creations. Hence, there is no predetermined future; it changes based on our choices every moment.

Only the results of our choices exist. We have the primary right to decide and shape our own lives. Those who work hard and succeed are incredible creators who have planted good causes, nurtured them, and reaped the results as concrete outcomes.

Types of Cause and Effect

Short-Term Causality: Situations where the result appears quickly.
(Example) Eating spoiled food and getting a stomachache in the evening.

Mid-Term Causality: Situations that take a few days to manifest, where a variety of factors play a complex role.
(Example) I applied for an interview at a tasty bakery but got rejected again. It seemed due to inadequate preparation for the interview, so I studied more to improve. Then, one perfect day, I got an unexpected interview offer and confidently succeeded. The place was a much larger French bakery than the one I initially wanted.

Long-Term Causality: Situations spanning a lifetime or even linking to past lives, composed of far more complex elements.
(Example) In elementary school, there was a boy, Jimin, who was ostracized for having only one arm. Despite the disapproval of my peers, I was kind to him. After moving schools in middle school, I lost touch with him but years later, while working on an exhibition, I discovered the invited artist was Jimin, who had overcome his disability and gained recognition for his artistry. He recognized me instantly and gifted me one of his most valuable paintings as a token of his gratitude for my kindness in our childhood.

Many people ignore these processes of causality, seeking only the results they imagine without effort. If you do nothing but fervently wish, will the universe fulfill your desires? Sorry, but that's just wishful thinking. Results without process are impossible, even for a deity. Results begin with a cause, and as various conditions align, the final outcome naturally forms. So, focus on creating causes, not results.

Getting upset and blaming others is futile. Most people blame others when they find reality unfair, denying responsibility for events that seem unrelated to them. This attitude is like trying to exempt oneself from the law of causality, akin to defying nature. Remember, even deities can't help with outcomes that defy the law of causality. Sitting idle and waiting for a miracle of reality creation won't work. To achieve results, fundamental actions must be taken in the physical world to create the right conditions.

As mentioned earlier, there are short-term causalities with immediate outcomes, mid-term causalities that take months or years, and long-term causalities that span lifetimes or even transcend into the next life. Depending on past karma, everyone starts at different points in life, each with unique challenges in this existence.

Sometimes people ask in dissatisfaction, "So, is it my fault that I was born into poverty?" My response is unequivocal: "Yes, it is."

There's no such thing as coincidence. Everything is necessity disguised as chance.

Every relationship and life circumstance unfolding before you is the result of a vast network of karma, even if it seems like a bad connection. We are born with everything pre-programmed—personality, physique, appearance, and life events. This analysis is the domain of disciplines like destiny science or astrology. Most people live according to the programs they're born with. This is partly due to the nature of causality and partly because it's challenging to make choices beyond our subconscious programming. Now you are aware of how powerful unconscious schemas can be, right? However, some people transcend their programming to create their own lives. They are the ones beyond all systems.

I have always been naturally drawn to spiritual journeys since my youth, leading to a keen interest in profound matters and encounters with related destinies. After first publishing *Zero* in Korean, I met one of our country's top destiny scholars. From him, I learned that everything is predetermined by one's natal chart, and nobody can escape this framework. Initially, this idea was met with resistance from me, as I had always advocated that human consciousness could create anything. However, my curiosity grew over time (in the early editions of *Zero*, I wrote that life is 65 percent predetermined, and 35 percent is subject to personal choice). I wanted to know if those who had turned their lives around were destined for such changes, or if they had transcended their predetermined lives through their own efforts. I became his student and studied traditional destiny science for several years, eventually giving it up for several reasons.

Firstly, I realized my intellect wasn't suited for understanding these complex systems (destiny science is analytical, which was difficult for my intuitive nature), and half-knowledge about someone's past, present, future, and relationships based on their natal chart could be as dangerous as an untrained shaman causing harm.

Secondly, I thoroughly grasped that everyone's life unfolds according to their destined fate. No one can escape the karma of cause and effect. My life was flowing exactly as determined by my karma, without any deviation! Even significant decisions in my life were moving along a predetermined path. Not only me but also famous personalities and even my destiny science master lived according to their fates, like reactions to the universal seasons. I was influenced by the flow of the moment I was born into. Perhaps being born into that flow was also part of my karma. But then I thought, if things are bound to happen anyway, what's the point in knowing beforehand? Knowing the future often brings unnecessary worry. Hence, I found it more helpful to encourage overcoming present challenges rather than locking reality in unnecessary fears about the future.

Thirdly, I discovered that there is one way to overcome one's programmed fate. Changing one's entire destiny is extremely difficult because everything happens due to causality. However, one can reduce the magnitude of events at critical crossroads in life or make the best choices through free will. But it's tough for an ordinary person to make choices beyond their inherent tendencies—maybe a 0.01 percent chance? It's not impossible, but not easy for everyone. Still, if someone can do it, they transcend their fate and can reshape and redefine their destiny. I decided to bet on nurturing this slim possibility.

Even if you know your future, events that must happen will occur, and no one can avoid them. If you desire to create or acquire something, you must create it according to your current situation. And karma is the cause that creates those real situations.

The core of reality creation is how well you resolve your karma, the skill of karma resolution. In this process, we understand the identity of the karma that governs us. If we realize we are in a programmed world, a slim chance emerges to become a "god," transcending the system. You can "choose" your program again with your free will.

"Bhikkhus, if someone says, "No matter what karma a person makes, they must experience its results," then there is no opportunity to cultivate wholesome actions and end suffering correctly. Bhikkhus, if someone says, "No matter what form of karma a person makes, they will experience its results," then there is an opportunity to cultivate wholesome actions and end suffering correctly."

—From the *Lonaphala Sutta* (AN 3.99, The Lump of Salt).

Rather than blaming the life circumstances we find ourselves in, let's discover the meaning and causality of our current lives. Pay attention to how we should react and what new choices we should make. If we focus on identifying and solving our current problems, this will become a new cause, leading to positive outcomes in the future.

Accepting life as it is, is crucial. Most dissatisfaction stems from the "rejection of life," an inability to accept. Once we recognize that all causes originate from us, our perspective on life will change. The Chinese phrase "結者解之" means that the person who tied the knot should be the one to untie it. Don't create negative energy by blaming impoverished parents or circumstances, which takes your mind further away from the Zero Point. Blaming implies that you have no fault, but that's not true. All causes of your life are within you. This life is tailor-made for us.

Some people born into poverty suffer and struggle, while others take it as a stepping stone and strive to live better. How we accept and respond to the life unfolding due to causality depends on our mindset. Our mindset is within our control. We always respond to phenomena, but it's up to us how we react. **The most important part of creating reality is how we perceive our lives, what mindset we hold, and how we view it.**

Recognizing that we chose to be born into such a family and remaining calm at the Zero Point to ask, "Why did this happen?" can lead to understanding life's causality and finding keys to solve problems. Such hidden connections need to be severed for "true" creation.

In reality creation, it's important to note that situations that feel like karma-induced challenges often arise. These variables can prevent situations from unfolding as we wish. The results of causes we planted unconsciously throughout our lives, some even from past lives, act as hidden variables.

When we wish for something, corresponding mental energy arises. The more intense the desire, the more powerful this energy spreads through the universe, attracting the right elements for creation. But this energy can also draw in unresolved causality that must be addressed first. Hence, new problems may surface before creation, which are these "hidden variables."

These variables can seem like obstacles, but they are also results of our own making. If we react angrily to these "obstacles," denying our role in creating them, we stray from the path of reality creation and become trapped in the system again. Remember, the mind should always remain in a Zero State.

The process of reality creation requires an ongoing task of recognizing and correctly discovering the causality unfolding before us every moment. I call this "reality reading." Reality reading involves identifying hidden variables that aren't easily visible and understanding their meanings to return to the Zero Point. By resolving these hidden variables through reality reading, conflicts and challenges that once seemed insurmountable can magically dissolve.

You have seen how the phenomenal world arises from the unmanifested or the causal realm. Reality reading is about observing and managing the changes in this causal realm. To do this effectively, one must first let go of oneself, as reality reading is not about subjective judgment or thoughts centered around "me," but about seeing through intuition.

Even though I've had magical moments in my life, there were also times of immense mental agony. One such time was during a relationship. Suddenly, the man I was dating faced a major crisis in his business and was extremely distressed. I was there to support him, thinking this was just a challenging phase. One day, he desperately asked me to lend him thirty million won ($23,000). It wasn't easy for me to lend such an amount since I wasn't financially abundant myself. But, out of love and a wish for his success, I lent him the money. Unfortunately, this was just the beginning. He continued to ask me for money, and I found myself unable to refuse. Eventually, I ended up paying off his debts, amounting to about 160 million won ($123,000), which led me to lose all my savings and even go into debt. I felt utterly unjust and incomprehensible. I wasn't at fault, yet I had to keep paying off his demands. Moreover, he neglected my struggles and even denied my existence. I felt like I was losing my mind.

I couldn't share my distress with anyone. I had paid his debts, but he flaunted his success to others as if it were his own achievements. Questioning him only brought me violence and torment. It was probably the darkest period of my life. I could understand why people with conditions like anger management issues or depression end up in such states. I lost my vibrancy and joy during those hellish years.

I pondered deeply about this inexplicable relationship. Why did this happen to me? What did this man mean to me, and how should I resolve this situation? Despite the warning from my destiny scholars teacher not to meet him, I foolishly believed he was my destined partner and faced tremendous hardships. These events coincided with a challenging period indicated in my destiny reading. It was a choice I made, and it was also a time when I was destined to face difficulties.

Did I get my money back? No, I didn't. Just being able to break up with him was a relief.

On the surface, it certainly seemed like it was his fault for burdening me with his problems. However, from a broader perspective, I realized that it wasn't solely his fault. Just like how life involves both earning and losing money, everything comes full circle, and I couldn't blame all the causality within my life's journey. Ultimately, I chose to release the anger and injustice I felt towards him, within me. Despite the significant harm he caused me, I decided not to harbor hatred towards him. I even chose to be grateful for the intense ordeal and darkness he put me through, as it became a crucial experience that allowed me, as a healer, to empathize more deeply with the pain of others.

Like my experience, we sometimes find ourselves in undesirable situations. In such times, you might proclaim the secrets of reality creation and strive to resolve the situation. However, that's not the only solution. First, you need the ability to accurately read why the situation occurred. These circumstances might seem like they're not the result of your actions. In my case, it seemed like it was all the other person's fault and I was just a victim. Sometimes, the losses we incur can seem utterly incomprehensible. But from a larger karmic perspective, things can look different. The money I lent to him might have been destined to be given to someone else. It could very well be that something within me caused the situation. In such cases, it's essential to look for the hidden variables of the larger causality, following the inner guidance.

Do you know what I learned from this incident?

I was very passive with money. I was good at saving money, penny by penny, but I was not good at spending it. Despite having a decent income compared to my peers because of my business, I had never spent a million won for myself. I was stingy with spending money. The thought of spending tens or hundreds of millions of won was unimaginable. When such an amount started to slip away all at once, I felt like I was going crazy. The loss I actually experienced was beyond imagination.

After this event, I realized that money I don't use has no meaning. I was someone who kept saving for the future but couldn't spend for the present. This incident made me realize that money exists to be used. I became someone who could give to myself and others. Losing a huge sum of money wasn't entirely negative. I became aware of my subconscious misconceptions about money and the lack thereof. By confronting and healing my subconscious lack surrounding money, I became freer from the notions that had bound me, and I could feel that the 160 million won ($123,000) I lost wasn't a waste at all. It was like tuition for life's lessons.

Of course, it took a long time to shake off this event. I had to understand situations that were otherwise incomprehensible and let go of resentment and hatred. Only when I released all these and returned to a state of calm zero, was I fully okay again. I realized once more that everything is a part of the complete journey of life and was grateful for resolving one of my lacks. With greater courage, I was confident that I could create a new reality again. My inner self already knew the reasons for that situation.

The ability to read reality is essential to understand problems arising from hidden variables and to achieve reality creation. Reality reading is essentially insight into life. Always practice looking at reality through intuition. The answers may lie in a place completely different from our thoughts. Your deep inner self will guide you to these answers.

workbook

How to Perform Reality Reading

To perform reality reading, a method for analyzing the fundamental causes of life's problems, we need to closely observe how we are responding to what's happening. Keeping in mind that all causes unfold from oneself, answer the following questions and try to view the situation in a multidimensional way:

1. What do I perceive as the problem in my reality?
2. Why do you think like number 1?
3. What do I feel about number 1 (how am I reacting)?
4. What were the things I was most afraid of or lacking in relation to number 1 before?
5. What experiences, sense of lack, or karmic predestination originating within me could have caused this situation as its result?
6. What can I learn from the same situation as number 1?
7. Were there any situations similar to number one before?
8. If all causes originate from me, what action will I take?

Tests of the Universe

I call the previously mentioned hidden variables, or life's obstacles, the Universe's Test. These are moments when we find ourselves entangled in difficult situations during life. Successfully navigating these tests is essential for us to progress to the next level. Perhaps, these moments are when you truly need your power of creation.

The Universe's Test is designed to mature humans and help them use their creative power correctly. It can be considered a mechanism of the universe. These tests are given in different forms, tailored to each individual's level. Everyone faces their own unique tests.

As one's creative power grows, the difficulty of these tests also increases. When one test is passed, subsequent challenges at the same level can be managed with relative ease. Therefore, when presented with a test from the universe, we should welcome and meticulously solve and submit our answers.

How can we recognize the Universe's Test? Sometimes, as if meticulously planned by someone, obstacles cunningly insert themselves into our lives, complicating matters. If the usual methods no longer yield answers, it can be considered a test from the universe. Isn't it better not to receive these tests? Absolutely not. If you settle for a comfortable life without obstacles, you also miss out on opportunities to enhance your creative power. So, when a test arrives, greet it with joy!

The Universe's Tests

— When I lose something I love, how well can I accept the pain as it is and let go of my attachment?
— When faced with something I dislike, how well can I accept that pain as it is and free myself from negative emotions?
— How free am I from past memories and information?
— Do I accept events as they unfold before me, or do I habitually rationalize them?
— Do I sincerely acknowledge that everything is the result of my choices?
— Am I creating more complex causal relationships through indiscriminate will and actions?
— Am I helping other creators, or am I causing them pain?
— Do I have the qualities to use the power of the universe?

The Hawaiian natives have a traditional healing method called ho'oponopono. This remarkable mechanism is based on the concept of healing others by purifying oneself. Traditional Hawaiian healers take the responsibility for the illness of the person in front of them and purify themselves genuinely to find and eliminate the cause of the patient's illness. We often blame our circumstances and view others in unfortunate situations with pity, separating ourselves from them. However, in *ho'oponopono*, the focus is on self-purification first. Why? Because self-purification is not just a simple healing mechanism but a transformative magic that changes all aspects of life. The world is constantly made up of opposing polarities and is given meaning by "consciousness" that observes them. All phenomena of life are merely mirrors reflecting ourselves. We perceive the world through what we experience and feel.

Do you want to know the real secret of this world? If I don't exist, the world doesn't exist either. The world exists only when I recognize this fact. If I don't perceive it, it doesn't exist. This idea might make you uncomfortable, but quantum physics has already stated that nothing exists before we become conscious of it.

Therefore, your perspective and thoughts are very important. What kind of world are you living in now? What world do you perceive and recognize? Is it a world full of sick people? A world dominated by evil? Or is it still a place with human warmth and kindness? Why do you perceive it that way? Could it be a world created by your thoughts?

In ho'oponopono, healing is done with the following four sentences and their hidden meanings:

"I'm sorry."—I'm sorry for causing your pain due to the cause I created.
"Please forgive me."—Therefore, please forgive me now.
"Thank you."—Thank you for giving me the opportunity to break free from negative causal relationships.
"I love you."—I love the universe and everything in it that has helped me grow through these tests.

It may seem trivial, but it creates a tremendous healing effect. Ho'oponopono became known to the world when a native doctor miraculously healed all the patients in a mental ward using this technique. By purifying only himself, he was able to heal everyone! It was possible because it was a 100 percent pure purification of how he defined and perceived himself and the world. The realization that others suffer due to the causes created by oneself, and fully accepting it, leads to the disappearance of all the karma and causal relationships that unfolded in one's life. When your vision, which perceived the world imperfectly, becomes clear, and you see a complete world, illnesses disappear miraculously. All this happens in "my" world!

I also reflect on the reasons behind the problems I face in life and seek out the unnoticed causal relationships, learning from them. During the eleven years I ran my business, I encountered numerous challenges. However, by examining the causes and zeroing myself, my fears never actually materialized. Maybe I was trapped in my fearful thoughts. Strangely, every time I broke free from the limits of my thoughts, the impending problems seemed to vanish instantly. As I resolved life's issues, I no longer perceived problems as problems. Do problems even exist? Were they just perceived as problems because I thought they were? Did they disappear when I stopped perceiving them as problems? I hope you find the answers through your experiences.

I had a client who came to me for help with a real-life problem. He was nearly a billion won ($770,000) in debt. He, too, had trusted someone, lending that person a large sum through a loan secured in his name. He was in a situation where he had to bear the interest of hundreds of millions of won every month. His business partner couldn't afford to pay the interest anymore, leaving him with the entire burden. He expressed his current situation and his desire to get a job. However, he was quite old and needed about 6 million won a month. Finding a job that paid that much was practically impossible, having already applied to over a hundred places without any response. He was terrified of the negative future looming ahead.

I decided to change his reality first. After calming him down, I advised him to look for happiness, getting out of the fear-filled thoughts. It was imperative to get him back to a Zero State. The reason for his predicament was his inner greed. He trusted someone else out of a desire for easy money and fame, lending a large sum in his name. It was all karma he had brought upon himself. I explained the reason for his reality, and though it was hard for him to understand at first, I didn't give up. I taught him to cleanse and understand his inner lack and desire, overcoming his problems step by step.

I reassured him that even though employment seems difficult now, creation can certainly happen, and he should not worry. I also made him specify the salary he wanted and the job he desired. I strongly told him to just believe that it could happen. He changed his mind and began to purify himself, gradually resolving the situation.

Miraculously, he received interview offers from several companies and went for interviews. I reminded him that the outcome was not yet decided, and he needed to continue keeping his mind at zero. Eventually, he got the job he wanted under the conditions he preferred. The negative future he feared never happened, and he managed to handle the situation by paying the appropriate interest. He also came to understand his own deficiencies through this ordeal.

Everyone can intuitively sense whether they have passed the universe's test through life's problems. We face countless tests from the universe. As we progress, the tests become more challenging. Sometimes, it feels like everything is being negated or lost, facing seemingly impossible situations. But don't be afraid. The universe's test is just a part of life. **Every problem is given only as much as you can handle and resolve. The key is not in "how to solve it" but in "how quickly you can escape fear." It may seem like what you fear will happen, but it doesn't.** First, come out of fear and look at the situation as it is. Then, start finding solutions within reality. The answer is always in reality. If you wisely find it, you have passed the test given to you. Life is like clearing stages in a game. By understanding and practicing the Zero System well, you can even pass high-level tests of the universe. Remember, everything in my life perceived as a problem is ultimately a test of the universe that I must overcome.

Awareness
— Self Information Correction System

By closely examining the jurors and schemas hidden in our subconscious and our automatic thoughts, we can discover numerous "errors" that we weren't aware of. While we easily spot others' mistakes, we are surprisingly lenient or in denial about our own. Therefore, it's crucial to check if our information processing system is functioning correctly.

If these errors are severe enough to manifest as psychological disorders, professional counseling is needed. But even if not, it's important to continuously self-examine and correct these errors to prevent worsening. Like our body's immune system defends against viruses, our mind has a self-correcting system to counteract "information viruses." Buddhism calls this awareness or mindfulness.

The ability to self-correct our information system is one of the greatest discoveries. Recognizing the problem accurately is the first step to solving it. Without a clear understanding of the problem, no method can change the situation. However, the subconscious jurors work hard to hinder this awareness, knowing that their errors would be auto-corrected, and leading to their deletion. The ego rebels to protect the "self," employing cunning defense mechanisms like projecting internal causes onto external objects to divert our attention from awareness.

The length of time we've lived with these errors is irrelevant because the mind always has the ability to reidentify itself from the origin. Anyone can objectively view their inner self, inspect it, and correct informational errors. Through this process, correct thinking becomes possible, allowing us to use our creative energy in the direction we truly desire, though it requires some effort.

Ways to Notice the Mind

1. When emotions arise, pause instead of reacting immediately. When an emotion comes up, don't get entangled but simply notice the fact that such an emotion has arisen in you.

2. Observe the emotion. Watch the emotion calmly, without reacting, and observe it as it is.

3. Name the emotion. Identify the emotion and give it the most accurate name possible.

4. Recognize where the emotion comes from. Once you've named the emotion, find out its origin. It's likely a response from a schema created by past experiences. It's just a remnant and adjunct of your memory, not something to identify with.

5. Notice that you've noticed the emotion. Realize that you've identified the source of the emotion and its impermanence. Once detached from the emotion, new thoughts and emotions might arise. Follow their flow and continue noticing.

6. Maintain awareness. Through the process of concentrating and practicing awareness, we realize this state can be sustained. Awareness leads to the realization that the initial thoughts and emotions were from subconscious errors, irrelevant to the present self. As you become more accustomed to awareness, you gain control over your mind, maintaining peace and reducing unnecessary arguments and conflicts.

Self-awareness of the subconscious may initially seem challenging, but as you start to notice each emotion one by one, you realize these are unnecessary in the present moment, mere residues of memories from the past. As you become accustomed to detaching from these emotions, which you once thought were "you," you'll begin to stop identifying with past emotions. Soon, these detached emotions will no longer draw energy from you and will dissipate.

Through this process, unnecessary emotions are eliminated, facilitating self-healing. Healing is not a grand power but a natural process of differentiating what's not "me" and separating it from the self, to rediscover the whole self. Self-healing is not a miracle reserved for the special few but a natural byproduct of routine self-monitoring and introspection.

"He's looking at you. Maybe he's looking at you because you're so fat!"

"How am I supposed to do this all at once? What if I fail? I'm so scared of failure!"

"Our subconscious continuously speaks to us, attempting to distort present information. Are you currently listening to the jurors of your subconscious? Remember, the true master is you, and you're also the employer of these jurors. "Knowing the enemy and knowing yourself, you can win a hundred battles without a single loss." What we need is just the right awareness. If the jurors interfere, calmly notice, "The jurors are interfering," and carefully examine the origins of their claims. Then, the jurors will realize they can't overcome the master and will disappear from the subconscious. We will then live truly autonomous lives.

Now, let's start paying attention to the deeper parts of our subconscious that we haven't noticed before. By looking into our subconscious and healing internal wounds, we untangle the knotted thread. You can't knit a sweater with tangled yarn; you must first untangle and neatly rewind it. As you become accustomed to introspection and identifying problems, you'll be able to transparently see reality without being controlled by the subconscious. Only then can we discover our true desires and convey them directly to the universe."

Collective Unconscious and Collective Karma

The world that one is born into is a phantom already present inside them from birth.
—Carl Jung

Not all causes in the subconscious are solely personal karma. The subconscious programming that constructs our notions and tendencies is also shaped by the collective unconscious of humanity, which includes unique cultural inheritances and experiences passed down through generations. The collective unconscious, as a vast pool of psychic energy, governs the consciousness of living individuals, continually generating powerful energy.

Korea is called a nation of *han* (deep-seated sorrow and resentment) due to the collective trauma stemming from thousands of years of foreign invasions and suffering. This trauma, although not experienced directly by individuals today, is inherited through the DNA of our ancestors. The increasing number of people in Korea who are like time bombs, ready to explode at the slightest provocation, might be due to this collective unconscious rooted in anger and *han.* The term *hwabyung,* a unique disorder to Koreans characterized by pent-up anger, was even listed in the DSM-4 (although removed in the DSM-5), indicating the concentrated energy of suppressed anger within Korean emotionality.

Different cultures form due to varying categories of the unconscious that are inherited regionally, shaping unique lifestyles. Just by being born in a country, individuals are influenced by its collective unconscious.

Individual consciousness is influenced by the collective. And one's fate intertwines with that of the group. No matter how much an individual strives, if war suddenly breaks out, they share the fate of their country, possibly needing to abandon everything to join the war or prepare to flee. All personal lives are affected by the collective karma.

For instance, consider yourself as a flower in a field. Your growth depends on the soil and environment. If the soil is fertile, you'll bloom beautifully; if barren, you might wither quickly. Additionally, the seasons play a role. In spring, you might flourish, but winter brings hardships. Some might be lucky to be planted in a pot and spend winter indoors, but most wildflowers endure tough times in the cold. Just as individuals have life seasons, so does the universe. Universal seasons affect everyone living in it, unrelated to personal karma.

The COVID-19 pandemic is an example. The virus itself isn't an individual problem, but perhaps a result of humanity's collective karma and unconscious, stemming from long-term environmental destruction. The entire world feels its impact.

The Doomsday Clock, managed by the Bulletin of Atomic Scientists (BAS), now indicates just one hundred seconds to "midnight," signaling humanity's precarious state. This situation results from human greed: disrupting natural orders for convenience and gain. Many of today's famines, environmental pollutions, natural disasters, and epidemics stem from human-induced ecological disruptions. These issues, threatening human survival, are a sort of global self-regulation against human "viruses" disrupting nature's program.

Our lives are intricately interconnected, influenced by collective unconscious and collective karma. Sometimes, we need to think and act in line with the larger community, as with global issues like COVID-19. To address such issues, a broader understanding of causality and an active approach are needed.

A famous case demonstrating the power of individual meditation in solving societal problems is the 1993 Washington D.C. meditation experiment. At the time, D.C. was one of the cities with the highest crime rates in the world. Transcendental Meditation practitioners predicted that meditation could reduce crime rates by over 20 percent. Despite skepticism, when the meditation began, crime rates started to decrease, with major crimes dropping by 23% when the number of meditators peaked.

Just as individuals bear the full responsibility for their lives, the collective responsibility of a society rests with its individual members. Sometimes, to resolve societal issues, we start with purifying the smallest unit—ourselves. After all, the concept of the collective exists because of the individuals within it. Isn't it amazing to realize that by purifying oneself, one can also bring about change in the larger society they belong to?

workbook

Purifying Collective Karma

The societal issues we face are also subjects for our embrace and purification, as we are the creators of these collective realities. Therefore, it's essential to take full responsibility for societal problems and send energy of care and love, even if they don't seem directly related to our lives. We can contribute to positive societal changes through self-purification and meditation.

1. What is currently the most pressing social issue?

2. What are the causes of step 1?

3. How am I reacting to step 1?

4. Write down any negative emotions I feel about step 1.

5. What within me has given rise to step 4?

6. Take all the responsibility upon myself, empty and purify myself.

7. Stay in a peaceful silence, and imagine filling myself and my surroundings with peace and love.

Transcend Karma

All desires originate from an assumption of lack. Therefore, every wish of mine not only manifests the desired outcome but also unwanted negative consequences. The reason why life often feels unsatisfactory is that we frequently face unwanted situations, but we must realize that these too are outcomes we have created ourselves.

We always hope for the realization of desired situations. However, if that "desire" stems not from a pure intention but from negative memories in our subconscious, it becomes difficult to attain satisfying results. Instead, life tends to become more entangled due to new cause-and-effect relationships.

We often seek external expert help when new problems arise. However, this effect is never lasting because other problems inevitably follow. We keep creating new problems for ourselves. Therefore, it is crucial to learn how to control and use the creative power within us.

Sometimes, we may not understand the root of our desires. In such cases, there is an alternative: not desiring. When you don't desire, the motivating mental energy does not exist, and hence, cause and effect are not created. "Not desiring anything" raises the question: then how do we create the life we truly want? The answer is "Desireless Desire." Desireless Desire refers to a wish without a reason, a wish not attached to outcomes, in essence, a pure wish. It may seem like a linguistic paradox, but the universe exists within such contradictions.

Desireless	Desire
Conscious Mind	Subconscious Mind

Most people desire many things driven by personal longing. These desires arise from a need to fill a void within. Ironically, these desires, driven by longing, may cause more harm than good in our lives. Desires rooted in longing can create unnecessary additional outcomes. The reason not to desire is that when the false self stops its superficial desiring, the path for the true self to follow becomes clear. This is the true purpose of our lives, arising from pure wishes within.

Desireless Desire is the genuine wish of the true self that emerges when the false self ceases its longing-driven desires. A pure wish transcends selfish motives, holding a pure intention for the good of all. It's not a wish for the personal self, but rather a wish in a state of egolessness. When Desireless Desire occurs, the universe begins to create reality according to our wishes. The creation from egolessness is fundamentally different from creation driven by specific thoughts.

To practice Desireless Desire, one must first drop all current conscious desires. These are likely not what you truly want. After stopping all desires, observe yourself deeply to find the accurate deep initial command. It's essential to not desire even this. Despite feeling contradictory, this is necessary to avoid disturbing the pendulum of universal relativity. Stop everything and let go. The deep initial command should be pure and simple, not born from lack. Then, transform this into a precise affirmation and place it in your subconscious. After this, don't desire anything further. Just stay at the Zero Point, maintaining a state of attunement with the cosmic consciousness. In this Zero State, where all thoughts cease, we connect with the universe, allowing us to draw the infinite energy of the universe into our reality.

Practicing Desireless Desire

1. Stop all current conscious desires completely.
2. Discover the deep, initial command.
3. Convert the initial command into an affirmation and just "recognize" it.
4. Let go of all intentions and stay at the Zero Point. Remain calm and tranquil, and then enter zero meditation.
5. Continue with everyday life, creating basic cause and effect, but don't strive to achieve it deliberately. The energy of non-action is already at work.

Process of Finding the Deep Initial Command

Once you've identified an initial command, keep asking "why" to delve deeper and discover the underlying, more profound initial command.

I want a red sports car.
→ Finding Initial Command: Is this truly what I desire? Why?
Modification: In reality, I need warm attention.
↓
I desire warm attention.
→ Finding Initial Command: Why do I want to be noticed?
Modification: I didn't receive much love from my mother as a child.
↓
I desire my mother's love.
→ Finding Initial Command: Why do I want my mother's love?
Modification: Well ... I just do.

The state of complete cessation of thoughts and desires, referred to as the Zero State, is akin to maintaining the pendulum's swing at zero constantly. It's an absolute state of tranquility where everything ceases—no wanting, no not-wanting. Yet, even in this seemingly void moment, the original desires we yearned for don't fade away. Our subconscious always remembers our deepest wishes, even when all thoughts cease. It's a state where everything stops, yet, paradoxically, the things we desire start happening. This is the essence of the Desireless Desire.

Imagine you realized your deep initial command for wanting a red sports car was actually a longing for your mother's love. When this longing is healed, astonishingly precise realities aligned with your deep needs start to materialize. You might even find the red sports car appearing before you. But your subconscious will be satisfied with any outcome that addresses the core need, whether it's a call from your mother or a new relationship. This happens because the deficiency in your subconscious has been addressed. Even if we are unaware of our root desires, it doesn't matter. When we cease all doing, our subconscious naturally guides us towards our desires. Just focus on stopping all thoughts and actions as much as possible. Miraculous events will start unfolding.

In a state where consciousness is connected with zero, things smoothly start resolving themselves. What seemed impossible effortlessly finds a solution, as there are no blockages in the flow of universal energy. This is especially true for desires that benefit others. Desires not solely for oneself but for others, offered purely, manifest easily. When blessings are sent to unspecified recipients, the energy directly acts without being affected by any causal relationships. As long as the material conditions are met, it promptly materializes.

However, since it's impossible to know all the causal relationships acting in our lives, it's crucial to be mindful not to add to them. Every action we take becomes a cause for another effect.

The only way to live beyond the law of karma (cause and effect) is to eliminate the "motive" behind it. Only then can our Desireless Desire transcend causality and become the purpose of our lives. Then, only joy from pure actions and experiences remains. If your intention behind helping the underprivileged is to be seen as a "good person," or if you're doing it out of pity, your actions are imbued with a specific motive. When the outcome differs from expectations, feelings of discomfort, anger, or betrayal arise.

For instance, let's imagine a kind lady who regularly donates to a beggar, feeling pity for him. One day, she sees him getting into a luxury car, which changes her perception and feelings. If her giving was truly selfless, the beggar's actions wouldn't have affected her. This illustrates the concept of zero energy—using energy without any attached desires or intentions.

Actions inspired by "desireless desire" are incredibly pure. But many people seek special meaning in their thoughts and actions. The lady's good deed was complete in itself. But why did she feel angry? What did she expect from the beggar? Was it not a relief to find he wasn't in dire straits? Maybe she derived comfort from his apparent misfortune, and its loss caused her distress.

Living a life of Desireless Desire and Actionless Action as its own end is truly beautiful, free from the projection of our false desires, causing no further cause and effect. Such a life naturally flows, and ironically, the happiness we deeply desired becomes realized, almost like a gift from the universe.

It's easier to fulfill Desireless Desires than to manifest intentions laden with desire, as the former transcends the law of cause and effect. We might exist to live this way. When we align our consciousness with the Zero Point and transcend causality, living each moment as its purpose, the universe willingly grants us what we need most. At the Zero Point, Desireless Desires naturally bloom into reality, bringing no worries or troubles. We learn that to be happy, we need not acquire or pay a price—we just trust ourselves and let go.

Desireless Desire isn't about mental energy but surrendering our life to the universe's natural order. This leads to encounters with blessings and creative moments beyond our imagination. By emptying our inner programming and becoming one with the universe, we can finally create and enjoy a life of abundance that we truly desired.

Isn't it simple? Just trust yourself and let go.

Just do it
Do it.

Zero Meditation

Zero Meditation is a meditation practice designed to place our consciousness at zero. It involves stopping all thoughts and simply dwelling in the empty space of zero. This state of consciousness is essential to master for creating reality. It's useful when there's mental unrest or when we need to set our consciousness to zero.

1. Sit comfortably.

2. Breathe in comfortably, and relax your breath as you exhale.

3. Zero Meditation is not about observing or noticing anything. It's purely for being present. Stop all motivations to do something and continue breathing comfortably without leaning towards any thoughts. However, this does not mean watching or being aware of the breath. Just feel the breath happening naturally.

4. The rule of this meditation is to stop the automatic perception and simply practice being. We habitually tend to perceive or think about something. Being caught up in any thought is also not zero. Thoughts belong to either the past or the future. Whenever a thought arises, step out of it and return to the empty space.

5. The point of zero is just an empty space. It's a state of completeness without doing anything. Just let the natural breath flow and stay in the state of zero.

The Process of Human Programming Reset

Shifting to the state of consciousness of Desireless Desire and the state of a Creator becomes possible when we reset our human programming. From birth, we have basic motivations, needs, and desires. If there's something we need to experience or achieve, we should just do it. However, for those who feel stuck or don't understand the purpose of their lives, it's essential to check their human programming and the structure of their conscious and subconscious information systems. This is because unnecessary information accumulates, preventing proper self-healing and reprogramming. To improve this state, we must reset our human programming, like formatting a computer.

The reset of human programming means resetting all our information systems. It involves discarding all the information and belief systems we have known and rebuilding them from scratch.

We have lived constructing an image of "me." This image might have been created by others or by ourselves. Our concepts form our unique frame, which, when solidified, becomes a stereotype that starts to standardize the world through our perspective. In fact, **errors occur due to such frames. Therefore, a reset process is needed to experience the realization of what remains when all things known as "me" are eliminated.**

When all the images I thought were "me" are discarded, what remains?

Who are you?

We start asking ourselves, "What am I?" When all conceptual systems are gone, what can we call "me"? By removing the "I," we experience a transpersonal state, a state of pure existence without any imposed information. Not a daughter of a family, not a student, not a woman, not a South Korean, not a high achiever, not a person with a cheerful personality ... only that indescribable state remains. Perhaps, that state is my true self. We experience a new perspective and world in this state of self-transcendence. (Repeated practice of the workbook "I Am I" in Chapter 2 will be helpful!)

Resetting human programming not only initializes our inherent program but also resets all artificial information we have lived with. It resets the information and limits of the binary world we lived in, and sets new information of the quantum transcendent world and the natural system of the infinite universe.

This is the start of new creation, the use of the complete energy of the universe, and the basic tuning process to live in harmony with the universe's flow. Without resetting human programming, we cannot use the system of the universe. Human programming is just a system programmed for our life. So, if we want to recreate our life, we first need to reset our system of consciousness. Like a full jar cannot hold new water, to receive new laws and energy, we must empty the old water and maintain an empty state.

The reset process of human programming progresses through a self-transcendence program that discards all images of "me." By formatting everything that could be called "me," we discover ourselves as a true being in a moment of transcendence. It's tough to discard the image of "me" on our own. A person or trained professional is needed to help. If professional help is not feasible, self-formatting through self-awareness can be attempted. Gradually working through the following workbook can help you learn the method of self-formatting.

workbook

Formatting Human Program 1

1. Observe the emotions that arise when experiencing things you dislike. Write down ten situations you dislike the most.

2. Choose one of the situations listed in step 1 and face it. Create or experience a similar situation and note down the emotions that arise

3. Separating from Unpleasant Emotions Using the "I Am I" Technique:
   Practice separating yourself from the unpleasant emotions that surfaced in step 2.
   — I feel displeasure/discomfort because of _________.
   — Is this feeling displeasing/discomforting "me"?
   — I am temporarily experiencing the feeling of _________.
   — What am I?
   — Quietly feel the "me" that is unaffected by any emotion.
   — Observe the emotion quietly. This emotion is
   not me. Where does this emotion come from?

4. Do the zero meditation.

By confronting situations that we dislike using the method above, we can delete the information related to the dislike and format ourselves so that these disliked things no longer affect us.

workbook

Formatting Human Program 2

1. Observe the emotions that arise when experiencing things you like. Write down ten situations you like the most.

2. Choose one of the situations listed in step 1 and face it. Create or experience a similar situation and note down the emotions that arise.

3. Separating from Unpleasant Emotions Using the "I Am I" Technique:
   Practice separating yourself from the unpleasant emotions that surfaced in step 2.
   — Do I like _________?
   — Is this feeling of liking _________ "me"?
   — I am temporarily experiencing the feeling of _________.
   — What am I?
   — Quietly feel the "me" that is unaffected by any emotion.
   — Observe the emotion quietly. This emotion is
   not me. Where does this emotion come from?

4. Do the zero meditation.

Continuously questioning why you like a particular thing leads to ambiguity in the original reasons for liking it. The certainty of why you liked something becomes unclear, and you may discover that there was no clear reason or you were feeling vicarious satisfaction for another reason. Recognize this and reset the rest in the same way.

workbook

Formatting Human Program 3

1. Consider what you want or desire and write down ten things.

2. Ask yourself if these are really what you want. Through continuous self-questioning, keep only what you truly desire and delete the rest.

3. Try to find the initial command in the remaining desires.
   — What I want:
   — Why do I want this? (First question)
   — Why do I want this? (Question about the first question)
   — Why do I want this? (Question about the second question)
   — Write down what you truly want, based on the answer to the third question. (Initial command)
   — Compare how different the found initial command is from what you initially wanted.

4. Identify the Lack (Relativity) in the Initial Command.

5. Do the zero meditation.

The things we want stem from a certain lack. This lack is an unfulfilled fragment of experiences in our lives. The lack can only be satisfied through experience. If you still find things you want to experience, think of ways to actually experience them.

workbook

Formatting Human Program 4

1. Write down ten beliefs you think are right.
2. Write down ten beliefs you think are wrong.
3. Write down ten beliefs you consider good.
4. Write down ten beliefs you consider evil.
5. Find and connect beliefs from the step 1 and step 2 lists that are related in terms of relativity.
6. Find and connect beliefs from the step 3 and step 4 lists that are related in terms of relativity.
7. Link similar beliefs from lists step 1 to step 4.
8. Review the connected beliefs and think about what kind of beliefs you predominantly hold.
9. What significance do these beliefs hold for you?
10. When were these beliefs first formed? Recall your memories and experiences.
11. Do the zero meditation.

workbook

Formatting Human Program 5

1. Write down ten things you think you must do.
2. Why do you think you must do these things?
3. What do you think will happen if you don't do these things?
4. What actually happens when you don't do these things?
5. What are you afraid of?
6. Do you have any similar past memories or experiences related to this fear?
7. What do you truly want? If what you truly desire is fulfilled, would you still want the things you listed in point 1?
8. Do the zero meditation.

workbook

Formatting Human Program 6

1. What are the things that concern you the most while doing the human program formatting?

2. If emotions like fear or doubt arise, write them down.

3. Likely, various defense mechanisms will trigger negative emotions. These emotions will disappear once they are experienced. Don't get caught up in the emotions—observe them calmly. Use the I am I technique to separate these emotions from yourself.
   — Am I feeling the emotion of __________?
   — Is that emotion of __________ "me"?
   — I am temporarily experiencing the emotion of __________.
   — What am I?
   — Let's feel the "me" that isn't swayed by any emotions.
   — Observe the emotion calmly. That emotion is not "me." Where does that emotion come from?

4. Do the zero meditation.

Defense mechanisms will arise whenever your schemas are triggered, hindering the reprogramming of your programming. You might feel like giving up, not wanting to face certain things, or feel doubtful. This is a very natural phenomenon. Don't be deceived by it; continue with the next workbook steps. Observe and separate arising emotions, and stay in the tranquil state of zero. As negative emotions gradually diminish, strong beliefs will start to soften.

workbook

Formatting Human Program 7

We will revisit the "I am I" exercise. This is a task to delete all illusions that I have known as "me" and to recognize my true self. I am a "me" that cannot be defined by anything. Practice erasing all the adjectives that make up me.

1. Write down all the words that describe you: (e.g., broke, tall, overweight, doctor, etc.)
   I am ________.

2. Look at each word in turn and consider if it really is you, or just an adjective describing you. If it's an adjective, try deleting it one by one and think about whether you can exist without it. If I can still exist without it, boldly erase each one. For example, "I thought I am broke, but is being broke me? Is it just an adjective describing me? Having or not having money isn't me. I was just mistaking the state of not having money for me." Then, erase "broke" and look at the next word, repeating the same process.

3. Revisit each word, discovering the self that mistook adjectives for"me." Then, feel the complete "me" with all adjectives gone. Once all deletion work is done, feel the existence of "I am I" with nothing else existing. No word in the world can define me. Place "me" where the adjective used to be and feel the complete self. I Am I.

4. Remember this feeling now. "I Am I." Feel the complete me with all adjectives removed.

5. Do the zero meditation.

workbook

Formatting Human Program 8

1. Reflect on the question, "Who am I?"
2. If any words come to mind that describe you, use the "I Am I" exercise to erase them.
3. You don't need to forcibly express it in words. Just close your eyes quietly and stay with the original feeling. Feel your true self that can't be expressed in any language. There's no need to define yourself. Just feel your existence
4. Continue naturally into a state of zero meditation.

After completing all eight steps, we return to a state of being reset, with no information left. When all information and concepts that could define "me" have been removed, we reach a state of transcending the self. In this state, I exist in my most pure and original form.

Now, let's start reprogramming with new concepts that can further expand us. The reprogramming process helps us reset to a state of consciousness as creators with infinite potential. Of course, these new concepts might not resonate with us immediately. However, if we enter the reprogramming with an open mind and a light heart, willing to expand our consciousness, it will surely help us view the world from a new perspective.

Reprogramming

The following are contents for reprogramming your consciousness. Read each one and meditate on its deeper meaning with your eyes closed. There's no need to think too hard or memorize anything. Just read each line lightly and open your mind to these new concepts.

1. There is no right or wrong in this world. The standard changes depending on the perspective. What we resonate with becomes right, and what we don't becomes wrong. Stop at the Zero Point of the relativity of right and wrong. Don't give power to either side. Everything is valid as it exists.

2. This universe is a space of infinite possibilities. Imagine as much as you can. The concept of "infinity" will start to form in you. Whenever you try to set limited boundaries for yourself, think of "infinity" again. Forget the limits you have had so far. The world starts to take shape as you perceive it. When you acknowledge infinity, infinite possibilities open up for you.

3. Once the concept of infinity is formed, you can use the power of the universe at any time. The universe exists within the world of infinity. Stay in the mind of the universe. Stay in the empty Zero State. Stay in the broader infinity where you disappear. I am the universe. I am everything.

4. If energy starts to lean too much to one side, realign yourself with the Zero Point. We are beings that can easily lean towards certain emotions. We might start to see the world through narrow thoughts again. If you lose your center, come back to the Zero Point. You now know where this Zero Point is. Return to your place of serene peace whenever you need.

5. To use the power of the universe, you must let go of artificial systems and follow the system of Wu wei (effortless action). Just observe the flow of nature. Look at the order of the world. Think of the most natural choice. Make fair choices for everyone, not for personal gain. When you resonate with the flow of Wu wei, the whole universe will assist you.

6. Everything is complete. Accept this completeness wholly. There

is nothing imperfect in this world. What you think is lacking is just your own deficiency. And even that is okay. You will soon climb onto the process of complete healing. Even if completeness feels far away, remember that nature and the universe were inherently complete.

When the concept of "I" disappears, we finally connect to the basic stage of the Zero State. Once reprogramming is completed, we are equipped with all the conditions to connect to the world of infinite creation. I wish you success and the best of luck in reprogramming!

Connecting to the Infinite

The Zero State is a state where we can freely use the infinite materials of the universe to create reality without any hindrance. To fully utilize the energy of the universe, our body and mind must be equipped with the right qualifications and conditions. The zero laws and exercises provided earlier are part of this preparation process.

It's only when we let go of the dichotomous thinking, fixed notions, and victim mentality instilled by society that we can truly realize our authentic selves.

Who are you? What can you do?

Does the world start to look different to you? In fact, the world hasn't changed at all. It's just your perspective that has shifted.

No matter how well-versed someone is in theory, if they haven't actually achieved what they desire, they can't understand the true meaning of "I am the master of the universe." However, those who experience and practice it firsthand can grasp its meaning. And when you truly become the master of your life, you can even understand the underlying principles and laws, just like I'm freely sharing about the zero system with you now. Don't forget what I've told you, and try to experience it gradually in reality. Through your own experiences, you will be able to verify the astonishing possibilities of reality creation. See for yourself whether I'm talking nonsense or if unbelievable things do indeed happen.

The next chapter is composed of content that we can directly experience and practice in reality. Now, you need to live as a creator of reality in your life. Knowing something in your head is not truly knowing it. Life isn't lived in the head, right? To truly know music, you need to go beyond reading the score and actually play and sing it, don't you?

Chapter 5. Right

The Life of a Reality Creator

Empty yourself. Let your mind be still.
Ten thousand things arise while "I" watch their return.
They grow, flourish, return, revert to their root.
Returning to the source is stillness, stillness is returning to one's nature.
Nature is constant.
Knowing constancy is called clarity.
Not knowing constancy creates reckless harm.
Knowing constancy, the mind opens.
Open mind leads to open heart.
Open heart leads to nobility.
Act with nobility, you will attain the divine.
Being divine, you follow the Tao.
At one with the Tao is eternal.
And though the body dies, the Tao remains.

— *Tao Te Ching*, Chapter 16

What is the Life of a Master?

You have learned about various laws of the universe and the meaning of a creative life. Yes, we all have the inherent right to be "creators," able to utilize the infinite energy of the universe. The only difference lies in whether we fully realize our creative power, whether we choose to be the master of our own life or not.

The infinite energy of the universe is open to everyone. However, if one does not realize this and choose to exist as a master, it remains merely a fairy tale in a distant land. Chapter 5 aims to show you how you can apply the zero system in your daily life and live a life of zero. As you reflect on your life and develop a habit of constantly being aware of the universe's zero system, you will soon experience an amazingly transformed life. The ability to create reality, like a gift from the universe, will be bestowed upon you.

Recognizing Things as They Are

We have never stopped thinking for a moment, nor have we lived without judgment. We are always thinking, and even when we try to stop, we are "thinking of stopping." Then, we find ourselves thinking, "It's not easy." Whether we are aware of it or not, we are living in the vortex of thoughts that follow one after the other.

As a result, this state becomes fixed and we fail to see reality as it is, getting trapped in our own framework and succumbing to distorted interpretations.

"I think, therefore I am."
—Descartes

When we look at a flower, we don't see it as it is. Instead, we recall past memories and think it's pretty or unattractive. Sometimes its scent can even feel unpleasant. When eating, our attention is often on the conversation or the TV screen, so we fail to fully taste the food.

We are accustomed to automatically judging everything based on information that comes to mind. Therefore, the original information conveyed by reality gets overshadowed, leaving us with reinterpreted and distorted information.

If only we could see everything as it truly is, we would discover and utilize new information within it. Perhaps we might even realize the essence of things that we hadn't noticed before.

When we observe objects or situations, we automatically go through the stages of perception → feeling → thinking. Therefore, emptying our thoughts is both difficult and important. Now, let's start practicing to empty our thoughts and see things as they are.

Through this practice, you will surely understand what intuition is, something I have been talking about all along.

workbook

Practicing Recognizing Things as They Are

1. Pick any object from your surroundings. Observe it and note down only the factual information you can gather from your observation. For instance, if you pick up a pencil, write down observable facts like: "It is yellow." "It has stripes." "There's an eraser attached." Try to note as many details as possible.

2. Once you can't think of more details to add, review the information you've written. Now, eliminate any subjective judgments or emotions included in your descriptions. Remove relative adjectives such as "sharp," "long," etc.

3. Look at the remaining sentences and observe the object again.

4. In the remaining sentences, mark a circle around information that seems common and obvious and a triangle next to information that you discovered or noticed anew today.

5. Count the number of erased information and the number of triangles you marked. This helps you realize how much you have been viewing objects through the lens of your preconceptions.

6. Repeat this exercise with different objects. Practice recognizing only the objective perceptual information, stripping away automatic emotional responses and judgments.

We often interpret objects based on our subjective thoughts without even realizing it. By setting aside these subjective thoughts and accepting only the objective information of objects, we can get closer to a Zero State and make intuitive decisions that often align surprisingly well with the essence of things.

To aid in observing things as they are, I'll introduce a helpful breathing meditation technique. This method, which involves observing your inhalations and exhalations and discovering the moment in between, enhances your ability to be mindful. Try practicing this regularly!

Inhalation-Exhalation Breathing Meditation Technique

1. Slowly exhale your breath.

2. When you can exhale no more, notice the moment of pause, the point where the breath turns.

3. Inhale slowly.

4. When you cannot inhale any further, notice the moment of pause again, the point where the breath turns.

5. Continue repeating this process, observing how your breath occurs. Notice the variations in the speed and depth of your breath throughout the cycle.

6. Realize that your breathing has never stopped since you were born. Pay close attention to the entire process of your breathing, which you might not have been fully aware of before, and observe how the breath happens.

7. If you find yourself distracted with thoughts like, "This is hard," "I can't concentrate," or, "I wish I could breathe more deeply," notice these thoughts and then gently return your focus to objectively observing your breath.

8. Recognize the "edge" where your attention to the breath wanes. When this happens, gently guide your attention back to observing the inhalation and exhalation. Maintain focus on observing your breath without any expectations or thoughts.

Tip: Training to Discover the Mind

The ability to transparently and honestly observe your mind as it is can be developed through practice. It doesn't require elaborate methods, but rather a straightforward approach of not deceiving oneself and simply observing what exists within. What's important is whether you are observing with an intention mixed within "yourself" or whether you are observing all intentions from a detached state, separate from "yourself." Discovering the mind always needs to be observed externally. Most of our conscious states are fixated on "self," making it difficult to distinguish whether our thoughts are those of the ego or the true self. This is why even with practice in discovering the mind, many people don't succeed in doing so properly.

Training to discover the mind should be done as often as possible, at every moment. Whenever you meet someone or encounter a situation, train yourself to become objective and observe the workings of your mind. This should always be done from the perspective of an external observer. Continuously refer to yourself by name and objectively view your own activities, as in "I, [Your Name], am having these thoughts right now!" As this practice continues, you'll begin to notice the transition from your current state of mind to the next. Then, whenever an undesirable state of mind arises, you'll be able to stop it in its tracks and replace it with different information. Real creative control over your reality becomes possible when you can observe in detail. This is why maintaining a state of wakeful awareness and the consciousness of an observer is essential.

Let Go of Desires

Desire is nothing extraordinary. Often perceived negatively, in its basic form, desire is merely a wishful heart. Wishing for something isn't inherently bad, is it? It's the driving force that propels us forward. However, when desires accumulate and take form, they start drawing out specific emotions based on past memories. These emotions are usually filled with unsatisfied desires or needs, leading to unwanted counter-reactions. Imagine being extremely hungry. Initially, you might just wish for something to eat, but as hunger intensifies, it brings other emotions like anger, potentially leading to undesirable situations. You might even resort to stealing food out of desperation. Desires can easily turn into greed.

Such desires can distort a lot of information and create difficulties as they flow in one direction. Therefore, we must be vigilant to not let our wishful hearts grow uncontrollably. When good things happen unexpectedly, without any specific desires, we refer to it as "unexpected fortune," felt more joyfully because it wasn't anticipated. Similarly, facing adverse situations without any specific desires allows for easier overcoming. Instead of despairing, these events can be viewed as tests from the universe, opportunities for new growth. Everything depends on your perspective and mindset.

The Zero Point is a state where neither good nor bad happens. We are conditioned to prefer good and reject bad through dichotomous thinking. To let go of this way of thinking, we need to let go of the discerning mind that separates good from bad.

When we are without specific intentions or desires, we can face all situations objectively. We identify with one aspect of a situation only because we desire something from it. If I say this, some people might start antagonizing the concept of desire. But remember, the emergence of desire is a natural phenomenon. We just need to be cautious not to let it grow strong enough to pull us away from the Zero Point. If you recognize a desire, don't judge it; just let it go.

Desire often stems from a sense of lack. However, this sense of lack is not who you are. You are complete as a creator, lacking nothing. So, no matter how burdensome the problems of reality may seem, don't fall into the folly of taking them as evidence of our incompleteness.

Experience the fullness and completeness of life in the Zero Point, even if it's just for a brief moment. The zero meditation I've shared will help. As the negative energy within us dissipates, you'll see that life's reality also starts to transform accordingly. This is the true secret to enriching life.

Live Righteously

Everyone's definition of "righteous living" varies due to diverse life experiences. However, in the Zero System, everyone's standard aligns, as a righteous life means maintaining balance without bias, living at the Zero Point.

When our thoughts and actions lean too far in one direction, imbalance ensues, creating surplus on one side and lack on the other. While one side experiences joy, the other endures sorrow. Like when one's victory joy leads to someone else's defeat sorrow.

Living righteously doesn't just mean living well. True righteousness transcends good and bad, finding the real standard in balance. Buddhism presents the Noble Eightfold Path as a practical way of life to attain enlightenment (a Zero State) by avoiding extremes and maintaining a middle path.

The Eightfold Path

1. **Right View**
2. **Right Intention**
3. **Right Speech**
4. **Right Action**
5. **Right Livelihood**
6. **Right Effort**
7. **Right Mindfulness**
8. **Right Concentration**

The Eightfold Path can be reinterpreted through the lens of the Zero System.

1. **Right View:** Understanding that the world results from causes we've created, continuously changing without permanent substance.

2. **Right Intention:** Not swayed by greed, obtaining pure information directly from reality.

3. **Right Speech:** Speaking in alignment with right view and right intention.

4. **Right Action:** Acting with Desireless Desires, making the act itself the

goal.

5. **Right Livelihood:** Earning a living without harming others or complicating cause and effect.

6. **Right Effort:** Doing one's best without attachment to results or rewards.

7. **Right Mindfulness:** Recognizing illusions as illusions.

8. **Right Concentration:** Focusing only on essentials, not swayed by illusions.

Reinterpreting the Eightfold Path in the context of the Zero System can provide fresh insights. These are not merely concepts to memorize but truths to be embodied through practice. Observing someone who seems to live rightly, they're likely aligning their life with these eight virtues.

While these principles might seem straightforward, actual adherence can be challenging. It requires continual self-emptying. The Eightfold Path is living wisdom that becomes evident through life's experiences. If practiced daily, maintaining balance becomes natural, keeping you anchored in the Zero Point.

Practicing the Eightfold Path liberates from skewed emotions, negative subconscious memories, and distorted information. It's a precious treasure of life, guiding us to dwell fully in the state of Zero.

Do Not Seek to Possess

Non-possession doesn't mean owning nothing; it signifies not holding onto the unnecessary. Our chosen simplicity, free from excess, is far more precious and noble than wealth.
—Buddhist monk Beopjeong

Humans always strive to possess something. Our desires and wants stem from this urge to own. We like to distinctly categorize things as "mine" and "yours." This ownership disease extends beyond mere objects and has long infiltrated deeper aspects of life. Parents believe their children belong to them, and husbands regard their wives as their property. Instead of respecting others as they are, people try to control them under labels like "my daughter," "my wife."

This is a grave misconception. Is there truly anything in this world that can be considered solely ours? Even our bodies are not truly our own. We say there is no self because the concept of "I" lacks substance. We cannot truly possess anything; we only form relationships with one another.

Many problems in life arise from attempting to own what cannot be possessed. Let go of the desire to own and see the precious people around you for who they truly are. You can never own another person. Instead, show love and maintain a close relationship. And to foster a healthy relationship, recognize the other as an equal being rather than treating them as a possession.

In this universe, all relationships are equal. The notion of one person dominating or owning another is, in reality, impossible. Even in human relationships, we should remain in a balanced zero state without bias. Only then can we truly see others for who they are and find freedom and flexibility in our own lives.

Nothing is Impossible

Our internal programming is often filled with restrictive and negative information, setting self-imposed limits. However, a few creative individuals have transcended these boundaries:

— People believed that humans couldn't fly, yet the Wright brothers invented the airplane.
— Helen Keller, unable to see, hear, or speak, overcame her disabilities and became a model of assertiveness and expression, graduating from college.
— Despite the belief that survival requires the consumption of five essential nutrients, there are individuals who have lived for years on water alone.

Flipping through the Guinness Book of Records, one might question the limits of human potential. These individuals seem to live in a reality so different from ours, almost as if they're from Mars. But they are human, just like us.

The difference between them and us lies in whether we accept our limitations or not. They embraced infinite possibilities instead of defining human limits. They actively believed that nothing is impossible or unreachable for them.

To initiate creation from Wu wei (effortless action), we must first position ourselves at the Zero Point. This Zero Point signifies a state where even our negative memories dissolve into nothingness. If negative memories and self-restrictions persist within us, they will inevitably manifest in our reality. When we attempt to create and doubt surfaces, wondering "Is this even possible?," it's a sign that negative energy is at play. With such a mindset, failure is inevitable. Forget all self-imposed limits, allowing only positive possibilities to remain.

Napoleon said, "The word 'impossible' is not in my dictionary."

Let's reset our internal programming with this new mindset:

There are no limits for me. I can become anything I desire. I have the power to realize it. Therefore, it will certainly come to fruition.

A limit only has meaning when we accept it. It is a self-imposed restriction. When we eliminate these negative limits, new creations start to unfold in our lives. From today, let's be a "Yes Man," affirming all possibilities and erasing the word "impossible" from our minds.

Just Make a Mark

Even if your goal seems to exceed limits, don't worry or overthink. Realizing it is not your task but the universe's. In the process of creating reality, all you need to do is make a "mark" representing your goal. You don't need to worry about how to reach it. The moment you mark it and focus on it, the universe naturally paves the way.

Consider the core principle of ski jumping: maintaining a balanced posture to fly far and land precisely. In ski jumping, it's incredibly challenging to control direction like walking on the ground. Ski jumpers rely on subtly adjusting their direction with the wind by entrusting their body to the air. Maintaining balance is a critical skill for them. Even a slight imbalance during flight can result in a failed landing or serious injury. How do they maintain balance and direction in the midst of nothingness? Ski jumpers direct their gaze towards their desired direction, not by turning their heads, but by focusing their consciousness there. Just the intent and subtle energy of looking in a certain direction can change their trajectory.

The principle of realizing dreams in life is similar. Like ski jumpers, we don't need to move our bodies or do anything specific. Our only task is to maintain the balance of mind and body. Maintaining balance doesn't mean staying in one place. Our life delicately aligns with the direction of our consciousness. Ski jumping, with its high velocity, requires much finer movements than on the ground. That's why jumpers need to enter a transcendent Zero State, where unnecessary actions don't exist. It's about entering a world where mental energy, not physical action, is crucial; a world where direction can be altered without moving the body.

Similarly, we don't need physical movement to head in a specific direction. Our only task is to remain precisely at the Zero Point. In the universal system, concepts of space and time don't limit us as the system exists beyond our perceptual boundaries. We believe we must constantly move to reach our goals, but in reality, our life's course is guided by finely focused consciousness—the Desireless Desire.

By staying at the Zero Point and marking our desired destination, the incredible power of the universe will lead us there. Of course, we can't do absolutely nothing. Minimal action is necessary to create life's cause and effect, but even that should be an intuitive action, a desireless action.

Relax all tension, and focus your entire consciousness on making that one mark. Don't try too hard to concentrate; it's just about making a mark. How hard can it be? The process of creating reality in the zero system is simple yet profound.

Wisdom Over Knowledge

We spend our lives learning and memorizing an abundance of information, labeled as "knowledge." In today's society, knowledge is often equated with "ability." However, the knowledge we firmly believe in is not always correct. History shows that widely accepted knowledge changes with time and culture, and new philosophies and scientific discoveries. Even moral standards and laws have been subject to change, haven't they? What was true in one era becomes a mere joke in the next.

"The only habitable planet in this universe is Earth."
"The body and mind are separate, and the body is just a precise machine."
"All diseases are a result of divine curse."

Knowledge is always evolving. What was right in the past may not be right now. Since we base our judgments on the information accumulated within us, if this information loses its relevance, serious errors can occur. Especially in today's rapidly changing digital world, it's often difficult to find answers with existing knowledge. So, what should we follow?

When unsure of what choice to make, pause all thoughts for a moment. Empty your mind and listen to your inner voice. Our pure inner self possesses the eye of wisdom. It sees situations correctly from all angles and continuously sends us signals. Trust your inner wisdom!

"I'm not sure why, but somehow this feels right."

There are moments when your heart moves, even if your mind can't understand it. Suddenly, you want to travel somewhere or do something, which doesn't fit your usual life pattern. It's important to note that our subconscious isn't always right. To obtain beneficial information from the subconscious, it must be separated from spontaneous desires.

At the Zero Point, without any interference, pure information arises. Following it leads to prosperity and happiness, and life flows smoothly in a direction supported by the universe. Pure information often emerges as thoughts or feelings that suddenly come to mind without any specific desire. When you follow these, things unfold in ways you never imagined—because it is the wisdom of the universe.

We already know everything. We just don't realize that we know it all.

Respect Everything

Respect means to highly regard and value something. We often show respect to those of higher status while neglecting those deemed lower. This stems from the customary notion that people of higher rank deserve respect from those below them, like a mother-in-law asserting dominance over a daughter-in-law, or an elder scolding a young person for not showing proper respect. In Korean culture, respect is often perceived as a courtesy owed only to those higher in status. Certainly, manners are necessary, but it should be a two-way street.

If energy flows only in one direction, it will eventually deplete unless it springs from an infinite source. Imagine if respect flowed both ways, creating an environment of mutual upliftment and harmony. When elders don't dismiss the young with, "What do they know about life?" and the young don't disregard the old thinking, "They don't understand the modern world," respect flows in both directions. This establishes a balance at Zero Point, where neither side is favored over the other. Everyone naturally wants to reciprocate respect when they receive it. This is a fundamental law of energy. The lesson here is simple: to receive respect, one must give respect.

This principle of respect is not limited to human relationships. Often, humans, considering themselves the most intelligent species, assume the right to dominate and exploit other life forms, leading to the destruction of habitats and loss of lives. Such anthropocentric views create imbalance in the energy flow, eventually causing ecological crises. We need to adopt a humble approach, respecting all forms of life. Earth's resources are not meant solely for human use. Coexisting harmoniously with all living things, humans can learn to use the universe's energy in a balanced way. To harness this zero energy, we must first align our mindset with zero.

By respecting everything, we can achieve a state of existence at Zero Point. Every form of energy then flows back to us, creating a life that is natural and harmonious. Everything, as it is, is perfect. All beings are equal; there is no lightness or heaviness in existence. Recognizing and humbly respecting the value of every being is all we need.

Our Life is Already Perfect

The systems of the universe are perfect, and as a part of it, we are complete beings. Can you believe that what seems an imperfect reality is actually flawless? If we escape the illusion of being inadequate and constantly recognize our completeness, imagine the transformation. Those who strive to stay at Zero Point through mindfulness suddenly realize one day that "Our life is already perfect. We lack nothing, desire nothing more, and everything is a blessing."

We have always lived a life of wanting. But when there's nothing left to desire, we are granted a complete life.

> On the seventh day, God had finished his work of creation, so he rested from all his work. And God blessed the seventh day and declared it holy because it was the day when he rested from all his work of creation.
>
> —Genesis, 1:2-3

Desires originating from lack can only be fulfilled by letting go of these desires.

When we see Michelangelo's perfect artwork, so flawless that it leaves us speechless in awe, we experience this completeness. The same applies when we look at our lives and the system of the universe. What more do we need? Everything is perfect as it is.

Everything in existence is complete. Look at the geometric beauty of a snowflake or the intricate structure of human DNA. No human can replicate the perfect system of creation and nature. We are already perfect beings just by existing.

When we realize this perfection and regain the meaning of our existence, we meet our true capabilities—the ability to use the infinite energy of the universe. We also reconnect with all perfect beings in the world. Only then do we re-enter the world of completeness. All the causes of our complaints disappear, allowing us to see the world anew, and live a harmonious and peaceful life.

You might still be skeptical, wondering how a life full of problems and setbacks can be perfect. But did a cynical approach ever change anything? You've always longed for more, never fully satisfied, returning to pain after brief happiness. If your life has been the same, why not flip it entirely and try living a life of zero? Let go of your mind of lack and believe in the unnecessary completeness and infinity within. Seek the true freedom of "nothing more to do."

Pause your rushing thoughts and actions, and begin meditating, looking within.

You have the right to live as you wish, with the ability to create a life of limitless possibilities. You are already a complete being. The moment you accept and choose to live as such, you can freely wield the power of creation. You couldn't use this power before because you didn't believe you had it.

Stop lamenting that life doesn't go your way. Don't despair over an incomplete life, generating desires from lack. Embrace your completeness and create endless possibilities with your existence. That is the present gift of the universe's system.

What "state of being" are you in now?

The essence of Zero Point is staying in the present. In the pure present, there are no emotions, memories, or information. Just experience this moment's empty life. No matter where you are or what you do, focus on and fully experience reality. Then, the "how" and "why" disappear. Just act!

If you're eating, focus on that moment. Savor the taste of the food and its colors. When walking, notice each step. Experience every muscle movement. Discover the many moments of life you've never noticed. Don't bring in unnecessary thoughts or emotions.

Just exist in this moment. Then you will experience a new state of being—a pure existence. Every moment, observe situations without discrimination as the only way to stay in the creator's state. In this space, there's no past or future, only the ever-changing present. And we just need to be aware of and exist in this constant change, in an eternal Zero Point. At this point, we meet our true selves and discover the real purpose and path of this life.

Indian mystic Osho said everyone can live life meditatively. True. We can choose to focus on and experience every moment of life. It's just a matter of choice, not exclusive to those with extensive training. Just choose to be mindful of what you're doing now, and you can enjoy every moment of life as meditation. Discover what you truly want to do and the meaning of life naturally.

We were born with all the answers. When we live as complete beings, the flow of the universe aligns with our true desires. Just like nature's phenomena follow a natural order, things start to unfold as we desire, without contradicting natural law. Then we become part of the flow, living as one with nature.

The moment we choose to live every moment as fully alive beings, we can enjoy all aspects of life. We can laugh when happy, cry when sad, be angry when upset, and be thankful when grateful. Without any past memories or information, only the pure feeling of the present exists. Doesn't it feel clean and clear? It's the state of purity, rediscovering the innocent smile of a newborn. And we will even be mindful of this fact.

Empty yourself to stay in the state of zero.
That is the path to awakening as a true creator.

This might be my last message to you.

I've tried to explain my experiences as clearly as possible for your understanding. The rest is up to you.

I hope you'll soon realize the full meaning of creating reality. If you truly grasp even one thing I've discussed, feel free to come back.

Farewell.

P.S. Here's a poem from my favorite poet that I wanted to share with you.

Love like you've never been hurt
—Alfred D. Souza

Love like you've never been hurt.
Dance like nobody's watching.
Love like you've never been hurt.
Sing like nobody's listening.
Work like you don't need money.
Live like today's your last day.

Singing Bowl Zero Meditation

Using a singing bowl, you can practice staying in a Zero State, a state of complete emptiness where thoughts cease. The point where the vibration of the singing bowl stops represents the absolute Zero Point. Let's try this simple meditation method to practice zero meditation.

1. Hold a singing bowl of suitable size in your hand.
2. Strike the singing bowl using a mallet.
3. Focus on when the sound of the singing bowl ends.
4. Once you identify the Zero Point (where the sound ends), stay in that moment of silence for about ten seconds.
5. Repeat this process, ringing the singing bowl and finding the end point. Try to extend the duration of the silent moment each time (ten seconds, twenty seconds, thirty seconds, etc.).
6. With practice, if you can remain in this absolute silence for longer periods, you will be able to access the Zero State anytime with just the sound of a singing bowl.

Pendulum Zero Meditation

The pendulum is an excellent tool for practicing emptying the mind. It swings in the direction of your thoughts, making it a useful aid for zero meditation. When asking the pendulum questions, it's essential to empty your mind and set it to zero, turning the use of the pendulum into an effective zero meditation.

1. Hold a pendulum in your hand. Keep your elbow supported on a surface so that it does not sway.
2. First, practice moving the pendulum using your mind. Think in your mind for the pendulum to swing vertically. Experience the pendulum moving vertically without physically moving your hand.

3. Think about the pendulum to move horizontally. Again, let the pendulum move horizontally solely based on your thought, without any hand movement.

4. Once the pendulum starts moving in the direction of your thoughts, imagine a direction and observe the pendulum move accordingly (e.g., spinning in circles, moving diagonally).

5. You've managed to move the pendulum with your will! Now, establish rules for directionality. Declare internally that if the answer to a question is "yes," the pendulum should move vertically, and if "no," it should move horizontally. This act is known as "programming." The pendulum will now move according to the rules you've set.

6. To meditate, prepare a deck of cards. Pick one card without seeing its face, so only the back is visible. Ask the pendulum a question about the card. For example, "Is this card black?" The answer will be either "yes" or "no," and the pendulum will start moving in one of the two directions.

7. This is the crucial part. The pendulum moves according to your thoughts. If you strongly believe the card is black, the pendulum will move vertically; if you believe it's red, it will move horizontally. It moves subtly according to your thoughts.

8. Watch the pendulum calmly and set your consciousness to zero. Empty your mind of any thoughts regarding the answer with the mindset, "I do not know." Focus more on calming your mind than on the purpose of asking the pendulum.

9. When you feel that your consciousness has reached a Zero State, then ask the pendulum your question.

10. The accuracy of the result depends on how well you maintain the Zero State. Repeatedly practicing this exercise will help you easily stay in the Zero State.

References

Ahn, Sangseop 안상섭. "Hangug-hyeong Ma-eumchaenggim Myeongsang-e Gibanhan Seuteureseu Gamso Peurogeuraem-i Manseong Tongjeung-e Michineun Hyogwa" 한국형 마음챙김 명상에 기반한 스트레스 감소 프로그램이 만성통증에 미치는 효과 [Effect of Korean-Style Mindfulness Meditation-Based Stress Reduction Program on Chronic Pain]. PhD diss., Yeungnam University Graduate School, 2007.

Assaraf, John, and Murray Smith. *The Answer.* Translated by Kyung-Sik Lee. Seoul: Random House Korea, 2008.

Baek, Jiyeon 백지연. "Sangdam Mit Simlijihoewa Wippasana Myeongsang-eseo-ui Al-ata-lim-e Gwanhan Yeongu" 상담 및 심리치료와 위빠사나 명상에서의 알아차림에 관한 연구 [A Study on Mindfulness in Counseling and Psychotherapy and Vipassana Meditation]. Master's thesis, Seoul Buddhist University Graduate School, 2004.

Begley, Sharon. *Train Your Mind, Change Your Brain.* Translated by Seong-Dong Lee et al. Seoul: Booksum, 2007.

Blanchard, E. B. "Stresses in Modern Life and Coping Strategies." In Mind, Machine, & Environment. Seoul: Hakmun Publishing, 1996.

Boorstein, Seymour. *Transpersonal Psychotherapy: Clinical Casebook.* Translated by Sung-Deok Jung and Luke C. Kim. Seoul: Seojo, 2001.

Choi, Beomseok 최범석. *Simsang Chiyoui Iron-gwa Silje* 심상치료의 이론과 실제 *[Theory and Practice of Mental Imagery Therapy].* Seoul: Sigma Plus, 2009.

Davies, Paul. *God and the New Physics.* Translated by Si-Hwa Ryu. Seoul: JeongShinSegyeSa, 1984.

Deikman, A. J. *The Observing Self-Mysticism and Psychotherapy.* Boston: Beacon Press, 2002.

Frawley, David. *Ayurveda and the Mind: The Healing of Consciousness.* Translated by Mi-Sook Jeong. Seoul: Shri Krishna Das Ashram, 1997.

Goldstein, Joseph. *The Experience of Insight.* Translated by Kum-Ju Lee. Seoul: Hangil, 1987.

Greene, Brian R. *The Elegant Universe.* Translated by Byeong-Cheol Park. Seoul: SeungSan, 1999.

Ha, Mijeong 하미정. "MBSR Peurogeuraem-i Jubu-ui Simlijeog An-yeonggam-e Michineun Hyogwa" MBSR 프로그램이 주부의 심리적 안녕감에 미치는 효과 [The Effect of MBSR Program on Housewives' Psychological Well-Being]. Master's thesis, Changwon National University Graduate School, 2008.

Hanh, Thich Nhat. *The Miracle of Mindfulness.* Boston: Beacon Press.

Hendler, Neal. "Depression Caused by Chronic Pain." Journal of Clinical Psychiatry 45 (1984): 30-36.

Herbert Benson, M.D., and William Proctor. *Beyond the Relaxation Response.* Translated by Hyun-Gap Jang, Ju-Young Jang, and Dae-Gon Kim. Seoul: Hakjisa, 1976.

Im, Eun 임은. *Han-uihak-gwa Yugyo Munhwa-ui Mannam* 한의학과 유교문화의 만남 *[The Encounter of Korean Medicine and Confucian Culture].* Translated by Jegon Moon 문제곤. Seoul: Yemoonseowon, 1999.

Jaeinyong 재인용. "Eunha Uju-do Changjo-han Jingong-ui Jeongche" 은하우주도 창조한 진공의 정체 [The Identity of the Vacuum that Created the Galactic Universe]. Wolgan Gwahang Newton 월간과학 뉴튼, 1998.

James, Oschman. *Energy Medicine in Therapeutics and Human Performance.* Translated by Young-Sul Kim. Seoul: Gunja Publishing.

Jang, Hyeongap 장현갑. *Ma-eum vs. Noe* 마음 vs. 뇌 *[Mind vs. Brain].* Seoul: Bulkwang, 2009.

Jeong, Eunhee 정은희. "Ma-eumchaenggim Myeongsang-i Am Hwanja-deul-ui Tongjeung-gwa Bul-an-e Michineun Hyogwa" 마음챙김 명상이 암 환자들의 통증과 불안에 미치는 효과 [The Effect of Mindfulness Meditation on Pain and Anxiety in Cancer Patients]. Master's thesis, Chungbuk National University Graduate School, Clinical Psychology, 2011.

Kabat-Zinn, Jon. *Healing Ourselves and the World through Mindfulness.* New York: Hyperion, 2005.

Kang, Giljeon, and Dalsoo Hong 강길전, 홍달수. *Yangja Uihak* 양자의학 *[Quantum Medicine].* Seoul: Wolgan Hwangyeong Nong-eop, 2007.

Kim, Seonggyu 김성규. *Ainsyutain-i Ggaedar-eun Yunhoe-ui Beobchik* 아인슈타인이 깨달은 윤회의 법칙 *[Einstein's Realization of the Law of Reincarnation].* Seoul: Keunsan, 1995.

Kim, Suwoong 김수웅. *Je5-ui Him* 제5의 힘 *[The Fifth Force]*. Seoul: Ullimsa, 2009.

King, Serge. *Mastering Your Hidden Self.* Translated by In-Jae Park. Seoul: Fragrance of Silence, 1985.

Laszlo, Ervin. *Science and the Reenchantment of the Cosmos.* Translated by Gyeong-Ok Byun. Seoul: Tree of Thought, 2006.

Lazarus, Richard S., and Susan Folkman. *Stress, Appraisal, and Coping.* New York: Springer, 1984.

Lee, Jongmun 이종문. "Uiryogigwan Jongsa-ja-ui Daech-eui Uihak Iyong Yang-sang Mit Insik-e Gwanhan Josa Yeongu" 의료기관 종사자의 대체의학 이용양상 및 인식에 관한 조사 연구 [A Survey Study on the Use and Perception of Alternative Medicine among Medical Institution Workers]. Master's thesis, Daejeon University Graduate School of Health and Sports, Department of Alternative Medicine, 2004.

Lee, Junseok, Byeonghwan Yang, Dongyeol Oh, and Giseong Kim 이준석, 양병환, 오동열, 김기성. "Juyo Uuljeung-eseo Uul-gwa Bul-an Jungsang-ui Sim-gakdo-e Ddal-eun Noepa A1, A2, Percent Bidaech-eung Jipyo-deul-ui Teugseong Yeongu" 주요우울증에서 우울과 불안 증상의 심각도에 따른 뇌파 A1, A2, percent 비대칭 지표들의 특성 연구 [Study on Characteristics of EEG A1, A2, Percent Asymmetry Indices by Severity of Depression and Anxiety Symptoms in Major Depressive Disorder]. 2006.

Lee, Kiseop 이기섭. "Padong Iron-e Geungeo-han Gi-ui Idong, Jeondal System Yeongu" 파동이론에 근거한 기의 이동, 전달 System 연구 [A Study on the Movement and Transmission System of Qi Based on Wave Theory]. Master's thesis, Wonkwang University Graduate School of Oriental Studies, 2009.

Lee, Kyungmin 이경민. "'Mu'wi" Gaenyeom-eul Tonghae Pyohyeon Doen Baram Imiji-e Gwanhan Yeongu: "Doing Nothing" Jeon-eul Jungsim-euro" "무위" 개념을 통해 표현된 바람이미지에 관한 연구 : "Doing Nothing" 전을 중심으로 [A Study on the Wind Image Expressed through the Concept of "Wu Wei": Focused on "Doing Nothing" Exhibition]. Master's thesis, Sangmyung University Graduate School of Arts and Design, 2005.

Lee, Kyung Sook 이경숙. *Ma-eum-ui Yeohaeng* 마음의 여행 *[Journey of the Mind]*. Seoul: Jeongsin Segyesa, 2003.

Lee, Seongkwon 이성권. *Jeongtong Gi Chiyubeop Sonbit Chiyu* 정통 氣 치유법 손빛치유 *[Traditional Qi Healing Hand Light Healing]*. Seoul: Health Digest, 2007.

Lee, Seunghun 이승헌. *Noehoheup* 뇌호흡 *[Brain Respiration].* Seoul: Hanmunhwa, 2011.

Lee, Taeseon 이태선. "Ma-eumchaenggim Myeongsang-i Yeogosaeng-ui Geungolgeog-gye Tongjeung-gam-e Michineun Hyogwa" 마음챙김 명상이 여고생의 근골격계 통증경감에 미치는 효과 [Effect of Mindfulness Meditation on Musculoskeletal Pain Relief in High School Girls]. Master's thesis, Duksung Women's University Graduate School, 2008.

Lee, Yooncheol 이윤철. *Jayeon Chiyu-wa Yangja Uihak (2)* 자연치유와 양자의학 (2) *[Natural Healing and Quantum Medicine (2)].* Seoul: Art House, 2009.

Mahasi Sayadaw 마하시 사야도. *Wippasana Myeongsang-ui Gicho* 위빠사나 명상의 기초 *[Foundations of Vipassana Meditation].* Seoul.

Mace, Chris. *Mindfulness and Mental Health.* Translated by Hee-Young Ahn. Seoul: Hakjisa, 2008.

Matthew Flickstein. *Journey to the Center.* Translated by Hyeong-Il Go et al. Seoul: Hakjisa, 1998.

Moon, Insu 문인수. "Daehaksaeng-ui Seuteureseu Daech-eobangsik-gwa Uul-i Eumju Haengdong-e Michineun Yeonghyang" 대학생의 스트레스 대처방식과 우울이 음주행동에 미치는 영향 [The Impact of College Students' Stress Coping Methods and Depression on Drinking Behavior]. Master's thesis, Konyang University, 2009.

Myowon 묘원. *12yeongi-wa Wippasana* 12연기와 위빠사나 *[Dependent Origination and Vipassana].* Seoul: Haengboghan Sup, 2006.

Nyanaponika Thera. *The Heart of Buddhist Meditation.* Translated by Wi-Ji Song. Seoul: Sigongsa, 1962.

Nyanaponika Thera. *The Power of Mindfulness.* San Francisco: Unity Press, 1972.

Okada, Kazuyoshi 岡田和佳. *Banya Sim-gyeong-gwa Saengmyeong-uihak* 반야심경과 생명의학 *[The Heart Sutra and Life Science].* Translated by Jang Sunyong 장순용. Seoul: Goryeowon.

Park, Sangwook 박상욱. "Noja-eseo Bon Muwijayeon-ui Gyoyuk Sasang" [Educational Philosophy of "Wu Wei" in Laozi]. Master's thesis, Kangnung-Wonju National University Graduate School of Education, 2009.

Park, Seok 박석. *Bakseok Gyosu-ui Myeongsang Cheheom Yeohaeng* (박석 교수의) 명상체험여행 *[Professor Park Seok's Meditation Experience Travel].* Seoul:

Mosak, 1998.

Park, Youngsook 박영숙. "Ma-eumchaenggim-e Gibanhan Seuteureseu Wan-wa(MBSR) Ung-yong Peurogeuraem-i Simlijeog Geongang-e Michineun Hyogwa - Satae Bul-an, Sahoe Simlijeog Geongang, Seuteureseu Daech-eobang-sik-eul Jungsim-euro" 마음챙김에 기반한 스트레스 완화(MBSR) 응용 프로그램이 심리적 건강에 미치는 효과 — 상태불안, 사회심리적 건강, 스트레스 대처방식을 중심으로 [The Effect of MBSR on Psychological Health – Focused on State Anxiety, Social Psychological Health, and Stress Coping Methods]. Master's thesis, Seoul Buddhist University Graduate School, 2006.

Pilowsky, I., C. R. Champman, and J. J. Bonica. "Pain, Depression and Illness Behavior in Pain Clinic Population." Pain 4 (1977): 183-192.

Pyeonjipbu 편집부. *Cheonjine* 천지인 *[Heaven, Earth, and Human].* Seoul: Hanmunhwa, 2009.

Romano, J. M., and J. A. Turner. "Chronic Pain and Depression: Does the Evidence Support a Relationship?" Psychological Bulletin 97 (1985): 18-34.

Sternbach, R. A. *Pain Patients: Traits and Treatment.* New York: Raven Press, 1974.

Shichida, Makoto シチダ マコト. *Cho Une Hyeokmyeong* 超ウ脳革命 *[Super Right Brain Revolution].* Tokyo.

Shichida, Makoto シチダ マコト. *Jineung-gwa Changjo-ui Jikgam-ryeok Gaebal* 지능과 창조의 직감력 개발 *[Developing Intelligence and Creative Intuition].* Tokyo: n.d.

Shin, Hyunkyun 신현균. "Sinchehwa Jiptan-ui Sinche Gamgak-e Daehan Haeseog, Churyeon Mit Gieog Pyeonhyang" 신체화 집단의 신체감각에 대한 해석, 추론 및 기억 편향 [Interpretation, Inference, and Memory Bias of Body Sensations in Somatization Group]. PhD diss., Seoul National University, 1998.

Taniguchi, Masaharu 다니구치 마사하루. *Saengmyeong-ui Silsang 1-gwon* 생명의 실상 1권 *[The Reality of Life, Volume 1].* Translated by Kim Haeryong 김해룡. Seoul: Hanguk Kyomunsa.

Talbot, Michael. *The Holographic Universe.* Translated by Gyun-Hyeong Lee. Seoul: JeongShinSegyeSa, 1992.

Washburn, Michael C. "Observations Relevant to a Unified Theory of Meditation." Journal of Transpersonal Psychology 1(10) (1978): 45-65.

Watson, Gay. *The Resonance of Emptiness.* London: Routledge Curzon, 2002.

Yoo, Yangsoo 유양수. *Gwahag-euro Barabon Summaek-gwa Geongang* 과학으로 바라본 수맥과 건강 *[Geopathic Stress and Health Seen through Science].* Seoul: Vision Publishing, 2008.

The author, Chunsia, whose real name is Jun Youn-kyung, has been acquiring the laws of the universe experientially through the practice of mindfulness since childhood. She has studied psychology, alternative medicine, Ayurveda, and meditation. Currently, she operates the Zentherapy Natural Healing Center, where she develops and manages programs that assist in people's consciousness growth and self-transcendence. She is also the innovator of the Singing Bowl Healing & Meditation System, a tool for utilizing consciousness based on the Zero Principle, and is recognized as a pioneer in establishing the culture of singing bowls in Korea.

* Bachelor of Psychology, authorized by the Minister of Education
* Master's in Psychotherapy, Graduate School of Alternative Medicine, Kyunggi University
* Ph.D. in the Department of Integrative Medicine at Sunmoon University
* Singing Bowl Master
* Director of Zentherapy Natural Healing Center, www.zentherapy.co.kr
* President of the Korean Singing Bowl Association, www.koreasingingbowl.com

Contact: keverei1@naver.com